India's Strategic Options in a Changing Cyberspace

India's Strategic Options in a Changing Cyberspace

Cherian Samuel
Munish Sharma

idsa
INSTITUTE FOR DEFENCE STUDIES & ANALYSES
NEW DELHI

PENTAGON PRESS LLP

India's Strategic Options in a Changing Cyberspace

Cherian Samuel and Munish Sharma

First Published in 2019

ISBN 978-93-86618-66-5

Published by
PENTAGON PRESS LLP
206, Peacock Lane, Shahpur Jat
New Delhi-110049
Phones: 011-64706243, 26491568
Telefax: 011-26490600
email: rajan@pentagonpress.in
website: www.pentagonpress.in

In association with
Institute for Defence Studies and Analyses
No. 1, Development Enclave,
New Delhi-110010
Phone: +91-11-26717983
Website: www.idsa.in

Printed at Aegean Offset Printers, Greater Noida, U.P.

Contents

Acknowledgements

The seed for this publication was planted in the final meeting of the Project Review & Steering Group when Brig. Abhimanyu Ghosh of the NSCS suggested that the various reports be compiled together as a book. This proved to be more difficult than expected, but we persisted, nonetheless, in the desire to add to the existing body of knowledge and debate on cybersecurity.

We would like to express our gratitude to the Ministry of Electronics and Information Technology for extending the grant to conduct the research. The Chairman and members of the Project Review & Steering Group – Mr. N. Sitaram, Brig. Ghosh, Dr. Ponnurangam Kumaraguru and Dr. P.S. Nageswara Rao – were generous with their time, providing extensive reviews, technical inputs, and constructive criticism. Former Dy. NSA Dr. Arvind Gupta and National Cyber Security Coordinator Dr. Gulshan Rai were constant sources of support and motivation; this book is only but a small contribution to their endeavours to expand research on cybersecurity within the country and making it part of the curriculum in academia and elsewhere.

The Institute for Defence Studies and Analyses (IDSA) has been supporting policy research on cybersecurity for a long time, and this book is a testimony to IDSA's continued commitment to policy relevant research and knowledge dissemination on security-related issues. The project, begun when Brig. Dahiya was at the helm of affairs, saw sustained support from Amb. Jayant Prasad when he took over as Director General IDSA, as well as Maj. Gen. Alok Deb, Deputy Director General. Gp. Capt. Ajey Lele fulfilled too many roles to mention, from sounding board to trouble shooter.

We are truly thankful to the anonymous reviewers of the book, and to all the individuals who have either motivated or supported the research, writing, and publishing of this book, both in their professional or personal capacities. Thanks are also due to the IDSA establishment and library, the team of

Mr. Rajan Arya at Pentagon Press, copy-editor Ms. Preeti Singh and Mr. Virender Negi for manuscript formatting. Additional project assistance was provided by Mr. Arul R. and Mr. Siddhartha Chandra Sekhar.

Last but not the least, our sincere gratitude to the entire phalanx of interviewees, speakers, panelists, and participants of the workshops, roundtables and conferences organized as part of the research project – Cmde. A. Anand, Brig. Abhimanyu Ghosh [Retd.], Mr. Abhishek Bansal, Mr. Amit Sharma, Mr. Anand Shankar, Amb. Asoke Kumar Mukerji, Ms. Bhavna Saxena, Ms. Chinmayi Arun, Mr. Deepak Maheshwari, Lt. Gen. Aditya Singh [Retd.], Maj. Gen. Ajeet Bajpai [Retd.], Mr. Gigi Joseph, Dr. Govind, Dr. Kamlesh Bajaj, Ms. Liis Vihul, Brig. Manjeet Singh, Mr. Mayank Lau, Mr. NandkumarSaravade, Mr. P.K. Agarwal, Mr. Rahul Sharma, Col. Rakesh Pandey, Ms. Rama Vedashree, Col. Sachin Burman [Retd.], Mr. Salman Waris, Mr. Sanjay Kumar Verma, Mr. SastryTumuluri, Ms. Satya Gupta, Ms. Shuchita Thapar, Dr. Sunil Agarwal, Mr. V. Anand Kumar and Mr. Vinayak Godse – who generously contributed of their time and shared their experiences in face-to-face interviews as well as through a series of interactions and roundtables, each one of which was invaluable in terms of the quality of inputs and the expansion of existing knowledge. Others, who have preferred to remain anonymous, are also duly acknowledged.

Cherian Samuel and Munish Sharma

Abbreviations

AES	Advanced Encryption Standard
APEC	Asia-Pacific Economic Cooperation
APT	Advanced Persistent Threat
ARF	ASEAN Regional Forum
ASEAN	Association of Southeast Asian Nations
CAC	Cyberspace Administration of China
C-DAC	Centre for Development of Advanced Computing
CERT	Computer Emergency Response Team
CI	Critical Infrastructure
CII	Critical Information Infrastructure
CISO	Chief Information Security Officer
COAI	Cellular Operators Association of India
CPNI	Centre for the Protection of National Infrastructure
DDoS	Distributed Denial of Service
DES	Data Encryption Standard
DHS	Department of Homeland Security
DIARA	Defence Information Assurance and Research Agency
DRDO	Defence Research and Development Organisation
DSCI	Data Security Council of India
DST	Department of Science and Technology
EU	European Union
ENISA	European Union Agency for Network and Information Security
GCCS	Global Conference on Cyber Space
GGE	Group of Governmental Experts
HQ-IDS	HQ-Integrated Defence Staff

IAMAI	Internet and Mobile Association of India
ICANN	Internet Corporation for Assigned Names and Numbers
ICT	Information and Communication Technology
IDSA	Institute of Defence Studies and Analyses
IDRBT	Institute for Development & Research in Banking Technology
ISAC	Information Sharing and Analysis Centre
ISO	International Organisation for Standardisation
IT	Information Technology
ITU	International Telecommunication Union
JWG	Joint Working Group
MoD	Ministry of Defence
NASSCOM	National Association of Software and Services Companies
NATO	North Atlantic Treaty Organisation
NCIIPC	National Critical Information Infrastructure Protection Centre
NIC	National Informatics Centre
NIST	National Institute of Standards and Technology
NSA	National Security Agency
NSCS	National Security Council Secretariat
OECD	Organisation for Economic Co-operation and Development
PLA	People's Liberation Army
PPP	Public-Private Partnership
R&D	Research and Development
S&T	Science and Technology
SCADA	Supervisory Control And Data Acquisition
SCO	Shanghai Cooperation Organisation
RAT	Remote Access Trojan
ReBIT	Reserve Bank Information Technology
STQC	Standardisation Testing and Quality Certification
UN	United Nations
USCYBERCOM	US Cyber Command

Introduction

As policymakers have found out to their consternation, cybersecurity is a moving target, with threats and actors manifesting and mutating at astounding speeds, so much so that even the most nimble governments are finding it difficult to fashion a viable response even within their own domestic domains. Various measures, ranging from developing Public-Private Partnerships, to classifying and concentrating on securing critical infrastructure, or devising offensive capabilities—all suggested by cybersecurity experts—have not been able to stymie the deluge of malicious threats emanating from state, non-state, and state-sponsored actors.

In the more innocent days of cyberspace, it was routinely referred to as a global commons, evoking visions of the various stakeholders coming together to develop norms and conventions to self-regulate their actions. The reasoning behind this was that this was the only way to further unleash innovation and creativity that had brought cyberspace into existence in the first place. Built on the principles of sharing, trust, and openness, security from malicious acts or software did not receive as much attention as it deserved. With rising economic costs and the national security implications of threats in and through cyberspace, calls for the rules of the road to regulate cyberspace have grown manifold. This has been sought to be achieved through the norms process wherein multiple stakeholders sat at the table to arrive at a consensus on various issues related to keeping cyberspace, "open, secure, stable, and free". This was largely a Western formulation, and it became the dominant framework by virtue of the fact that much of the technology and capacity as well as private enterprises underpinning cyberspace resided in the Western world. Norm negotiations have straddled various forums in the multilateral, multi-stakeholder, technical, and functional spheres. The complexity of the issues to be negotiated, and the differing approaches and perspectives have increasingly made it difficult for these negotiations to continue at a meaningful pace. Differences of opinion related to, *inter alia*, governance aspects, the responsible behaviour of States, militarisation, legal obligations, the right to self-defence, etc., are just some of the impediments that have come in the way of the process of norms building.

More and more countries have begun to assess their vulnerabilities to cyber attacks, and are exploring all ways and means to deter and respond to such attacks. The changes in military strategies and doctrines factoring the need to raise cyber commands or strategic forces reflects the prominence nation states are now giving to building a deterrent force or posture in cyberspace. States are not reluctant anymore to develop and practice offensive cyber capabilities as part of strategy to secure their interests in cyberspace. There is increasing international scholarship and debates about applying deterrence and its associated concepts to cyberspace given the vital role the concept has played ever since World War II. Adapting deterrence to cyberspace has taken root in the absence of any other viable conceptual framework to work out a strategy for responding to attacks. Both these frameworks are preferred options, partly because they ascribe roles for States.

This also underscores the increasing salience of cyberspace, both as a strategic domain for operations as well as an enabler or force multiplier for effects in and through other physical domains, namely, land, air, space, and water. That said, there is no dearth of evolving concepts, emerging perspectives, and novel dimensions of technology to address the critical issues pertaining to the security, safety and stability of cyberspace. The concept of Active Cyber Defence, incorporating classic military techniques—such as the kill-chain—and adapting them to cybersecurity requirements is an example. However, it is yet to gain much traction despite much discussion and debate, largely because of the complex legal and technical issues involved.

Technology is a key vector, acting to both the benefit and detriment of cyberspace. Information and Communication Technologies (ICT) have helped billions of people to access the Internet with ever increasing penetration and ubiquity; it has been an enabler, a facilitator, and a provider of opportunities in terms of jobs, businesses, ideas, and innovation. However, in the post-World War II era, technology has been at the forefront of strategic competition, be it the race to develop critical technologies (most recently in domains like artificial intelligence and quantum computing), and then to both showcase this technological prowess as well as to deny it to others through obstacles like stringent export control regimes for technology control and denial or by creating proprietary standards. Be it old technologies (like encryption) or emerging technologies (like quantum computing and artificial intelligence), policy makers are faced with the dilemma of fashioning rules and regulations that keep in check the negative aspects while allowing space for innovation.

Encryption, the ancient practice of hiding secrets, has been a subject of strategic interest and subject to stringent export controls right since the end of the World War II, given its applications in protecting secrets and communications for the military and for governments. In the quest for secure information and communications in cyberspace, encryption has emerged as one of the promising technology solutions. It is, perhaps, also a potent factor in the race for technology dominance. Encryption, whether in the form of digital signatures or public-key cryptography or certificates, forms the bedrock of the digital economy, secure banking, and burgeoning e-commerce. But, privacy concerns, in the backdrop of mass surveillance especially by democratic governments to provide security against crime and acts of terrorism, have fuelled the debates over government's lawful access to encrypted data. Led by civil society and privacy advocacies, the debate impacts billions of people across the globe, and the respective governments are formulating their own laws and mechanisms to address this, in accordance with domestic limitations and interpretations of civil liberties.

As cyberspace expands and threat actors acquire more tools, secure communications have acquired centrality, and propelled the quest for novel technology solutions. There is a continuous race between cryptography and cryptanalysis—the ability to produce relatively secure cryptosystems and the techniques to break or undermine those cryptosystems. Quantum computing, for instance, may render some prominent encryption standards either vulnerable or insecure. Technological answers to the "Post-Quantum" phase of information security are being pursued aggressively all across the globe. In other words, no country is immune to the dynamic changes cyberspace has brought in to the practices of foreign and military policy making, technology development, or even to the protection of their critical information infrastructure upon which the modern nation states are heavily dependent.

The vulnerabilities of Critical Information Infrastructure are particularly troublesome. They are under constant threat, owing to underlying vulnerabilities woven with the legacy technologies, their vast geographical spread, and for geopolitical reasons, the interest of nation states in exploiting them, whether it is a nuclear installation, an electricity grid, or a government database storing personal information. System failures are the virtual equivalent of natural disasters but the responsibilities to secure the information systems, cyber assets, and in turn cyberspace, are not strictly that of governments alone. They require equal effort on the part of both the public and private sectors. Securing cyberspace opens immense possibilities for Public-Private Partnerships, at both strategic and operational levels, with the private sector providing invaluable technical and managerial expertise to augment governmental efforts.

India has high stakes in the security and stability of cyberspace, not least because the government has made digitisation a priority, and is in the process of executing national flagship programs to improve governance and the delivery of essential services, which also includes securing world's largest database holding the personal information of a billion plus Indians. Critical infrastructure, which underpins our economic and social well-being, has a lot more dependence on cyberspace, and any compromise may have a debilitating impact on the national economy and national security. In addition to security imperatives, cyber technologies have the potential to propel India's development as a knowledge economy and the next generation of economic growth.

Providing thought leadership in cybersecurity requires both clarity of vision, and an understanding of the history and evolution of cybersecurity. As India aspires to find a place on the policy high table of cyberspace governance to play a prominent role, thought leadership and technology innovation are some of the attributes which will embellish India's credentials. This book is an attempt to unravel the vast changes in the processes of norms building, the emerging concepts shaping military thinking as well as the protection of critical infrastructure and the evolving technology sphere pertaining to cyberspace, as also to gauge India's position. It is compiled out of a two year project sponsored by the Ministry of Electronics and Information Technology (MeitY) and is based upon research conducted by the authors as well as by invaluable inputs from various seminars, round table discussions, workshops, interviews, and reports compiled during the course of the project. Whilst the project comprised of standalone reports, these have been modified and adapted to provide continuity to the contents of the book. It is structured to perform a role both as a primer to those who wish to understand the strategic issues and key concepts in cyberspace, as well as to provide sufficient pointers to those who wish to have an in-depth understanding on specific issues.

The authors have made their best efforts to remain factually correct, and they solely are responsible for any errors. The views expressed are personal, and do not necessarily reflect the views of IDSA or the Government of India.

CHAPTER 1
Concepts and Definitions

Cyberspace

The quest to develop a common terminology for cyberspace is an ongoing one, and is seen as an important precursor to developing norms for cyberspace in the medium term, and treaty agreements in the long term. In its June 2010 Report, the United Nations Group of Governmental Experts on Developments in the field of Information and Telecommunications in the context of International Security (UNGGE), recommended "(v) Finding possibilities to elaborate common terms and definitions relevant to General Assembly resolution 64/25."[1] Though specialising in cybersecurity, the name of the group belies its origins (in 1998) as an initiative of the Russian government which proposed the establishment of the Group of Governmental Experts (GGE) to examine the issue of information security. This basic conflict over what constitutes cyberspace, cybersecurity, and other related terms is yet to be resolved, with the number of alternative definitions increasing by the year.

The US Government's National Strategy to Secure Cyberspace, released in 2003, defined it as follows: "Cyberspace is composed of hundreds of thousands of interconnected computers, servers, routers, switches, and fibre optic cables that allow our critical infrastructures to work."[2] The focus of the definitions varies according to the nature of the parties involved in the drafting of the definition. The US military, for instance, defines cyberspace as "a global domain within the information environment consisting of the interdependent network of information technology infrastructures, including the Internet, telecommunications networks, computer systems, and embedded processors and controllers." As early as 1995, the US military had described Information

Operations as the fifth dimension or domain of warfare after land, sea, air, and space.[3] By 2008, the definition had expanded considerably: it was now defined as "the interdependent network of information, technology infrastructures that includes the Internet, telecommunications networks, computers, information or communication systems, networks, and embedded processors and controllers in critical industries."[4]

The Russians say that they do not have a word for 'cyber' in the Russian language; the closest approximation is 'information' which includes not only data but even the thoughts in one's head.[5] A US-Russian project to come up with definitions of critical terms in cyberspace could only agree on the bare minimum, describing cyberspace as an "electronic medium through which information is created, transmitted, received, stored, processed, and deleted."[6]

The National Cybersecurity Policy of India - 2013 did not attempt to create a definition but used the definition found in the International Standards Organisation Guidelines for Cybersecurity released in 2012. This states that cyberspace is "a complex environment consisting of interactions between people, software services supported by worldwide distribution of information and communication technology."[7] The International Standards Organisation gave a caveat to this definition, noting that "there are security issues that are not covered by current information security, Internet security, network security, and ICT security best practices as there are gaps between these domains, as well as a lack of communication between organizations and providers in the Cyberspace."[8]

Cybersecurity

Cybersecurity is an equally problematic term when it comes to definitions, and much of the problems arise from the complexity of cyberspace. As the International Standards Organisation (ISO) document notes,

> the devices and connected networks that have supported Cyberspace have multiple owners, each with their own business, operational, and regulatory concerns. The different focus placed by each organization and provider in Cyberspace on relevant security domains where little or no input is taken from another organization or provider has resulted in a fragmented state of security for Cyberspace.[9]

The differing focus can be seen in definitions of cybersecurity as proposed by the three crucial sectors: commercial, defence, and civil society. The definition proposed by the International Telecommunications Union (ITU) is indicative of the first. It states:

> Cybersecurity is the collection of tools, policies, security concepts, security safeguards, guidelines, risk management approaches, actions, training, best practices, assurance and technologies that can be used to protect the cyber environment and organization and user's assets. Organization and user's assets include connected computing devices, personnel, infrastructure, applications, services, telecommunications systems, and the totality of transmitted and/or stored information in the cyber environment. Cybersecurity strives to ensure the attainment and maintenance of the security properties of the organization and user's assets against relevant security risks in the cyber environment. The general security objectives comprise the following:
>
> Availability
>
> Integrity, which may include authenticity and non-repudiation
>
> Confidentiality

An example of the second is the extended definition put forward in the US Department of Homeland Security's glossary of Cybersecurity Terminology:

> Strategy, policy, and standards regarding the security of and operations in cyberspace, and encompass[ing] the full range of threat reduction, vulnerability reduction, deterrence, international engagement, incident response, resiliency, and recovery policies and activities, including computer network operations, information assurance, law enforcement, diplomacy, military, and intelligence missions as they relate to the security and stability of the global information and communications infrastructure.[10]

A final example is that put forward by the Freedom Online, with a focus on human rights, "Cybersecurity is the preservation—through policy, technology, and education—of the availability, confidentiality and integrity of information and its underlying infrastructure so as to enhance the security of persons both online and offline."[11]

Cyber War

Though there is no one definitive definition of cyber war, the most-quoted version is that by Richard Clark and Robert Knake in their 2010 book *Cyberwar* which defined cyber war as the

> unauthorized penetration by, on behalf of, or in support of, a government into another nation's computer or network, or any other activity affecting a computer system, in which the purpose is to add, alter, or falsify data, or cause the disruption of or damage to computer, or network device, or the objects a computer system controls.[12]

One of the revelations of the UN GGE's recommendation that international law and the UN Charter apply in cyberspace was that the only international laws found relevant to cyberspace were the Laws of Armed Conflict and International Humanitarian law. The consequent focus on cyber war was further aided by the release of the Tallinn manual on the applicability of these laws to cyberspace. The articles of the UN Charter relevant to cyber conflict included Article 1 on maintaining international peace and security, Article 2(4) on the use of force, and Article 51 on the right to self-defence. Article 2(4) of the Charter states that "[all] members shall refrain in their international relations from the threat or use of force against the territorial integrity or political independence of any state, or in any other manner inconsistent with the Purposes of the United Nations."[13] At the time of the framing of the UN Charter, the term 'use of force' was kept ambiguous, though the developing countries wanted 'use of force' to cover economic as well, and to be included during the drafting of the Charter.

The developed countries were against the inclusion of more precise terminology since they saw economic and political coercion—such as economic sanctions—as a useful tool of diplomacy.[14] In the cyber context, the Charter does not elaborate on use of force under Article 2, or the meaning of armed attack. Subsequently, in 1974, the term 'aggression' was defined through a resolution of the General Assembly as follows: "Aggression is the use of armed force by a State against the sovereignty, territorial integrity or political independence of another State, or in any other manner inconsistent with the Charter of the United Nations, as set out in this definition."[15] Here, too, the caveat was that, in order to merit a response at the United Nations, it had to be of sufficient gravity in order to be condemned.

Cyber Weapons

One of the most difficult issues related to adapting the existing international law to cyberspace is to do with cyber weapons. There is, as yet, no legally agreed upon definition of a weapon, and the unique characteristics of cyberspace makes defining a cyber weapon that much harder. This raises the legitimate question as to whether cyber weapons are a reality, even though there are ample examples of so-called weaponised malware like Stuxnet, Flame and Duqu.

Conventional Weapons have been classified and regulated on the basis of their ability to kill, injure, or disable people, or cause significant damage or destruction to property. Those that are deemed to cause immeasurable suffering, such as chemical and biological weapons, have been sought to be banned through conventions. Attempts at regulating weapons go back to the St. Petersburg Declaration of 1868 which banned the use of projectiles of less than 400 grams. Going by such classifications, none of the malware can be classified as cyber weapons since they do not demonstrably kill, injure, or disable people, and when property is damaged or destroyed, it has happened below a threshold considered as tantamount to an aggressive action. The two main areas where a definition of cyber weapons would help are, i) in adapting the laws of armed conflict to cyberspace, and ii) in considering cyber arms conventions or treaties.

The laws of armed conflict are not so much concerned with defining weapons as they are with reducing human suffering in conflict. The same applies to international humanitarian law. To the extent that suffering is caused by conflict, principles such as proportionality, distinction, and necessity are used to regulate the use of force. Under the principle of necessity, "forces must engage only in those actions necessary to achieve legitimate military objectives"; they must distinguish between lawful and unlawful targets, such as civilians and civilian property as well as innocent third parties. Proportionality refers to prohibition on the excessive use of force. Given the wide range of malicious activities in cyberspace, adapting these to fit precise legal frameworks would be a difficult task, compounded by the difficulties of attribution further complicating the adaptation of these principles to cyberspace.

One line of thought is that cyber weapons, in fact, lessen human suffering since they are bloodless weapons, and suffering in its classic sense of causing injury or death does not take place. Damage and destruction to property can and does take place, but not on the scale envisaged in the laws of armed conflict.

Any cyber arms control treaty or convention would also require definitions and the classification of cyber weapons to be effective and enforceable. Classic arms control treaties restrict and control the spread of physical weapons through regulation and compliance. This is difficult with cyber weapons because, they have unique attributes including dual use, ease of replicability, anonymity, lack of attributability, and difficulty in monitoring and verification.

Arms control treaties not only seek to restrict the weaponry but also to restrict the spread of the underlying technologies. This is difficult in the case of cyber weapons because of some of the attributes mentioned above. Once used, the weaponry, the methodology—indeed everything is available for analysis by everybody; it can be replicated and modified to suit different purposes, as seen in the case of Stuxnet.

Other reasons why arms control treaties are seen as a long term objective is that the technologies are still maturing, and restrictions would constrict their development. On the flipside, it makes sense to lay out restrictions while the technology is in development and not after, which could result in some countries having access to the technology, as in the case of the Nuclear Non-Proliferation Treaty.

This also explains why countries that have been a victim of weaponized cyber attacks, who should ideally be in the forefront of demanding bans on such weapons, have not been vocal. They also wish to develop cyber weapons capabilities without being constricted by restrictions. The downside to this is that, i) cyber criminals have more and more sophisticated malwares at their disposal, and are having a field day in the absence of any international agreement on malicious activities in cyberspace; ii) cyber espionage has been considered an acceptable activity even though most of the malware used is multi-purpose in that it can be used to destroy, degrade, exploit, control, deceive, and alter devices on a network. Rather than splitting hairs about different forms of cyber espionage, it might be time to consider the norms for cyber espionage. It is not only States that are spying on each other, but even companies are spying on companies, and companies on States, and States on companies. The multiplicity of actors increases the chances of misperception and escalation; iii) with cyber weapons largely being developed by the intelligence agencies and the military, the political leadership is largely in the dark about the weapons and their capabilities. As capabilities increase, discussions at a political level are needed on the issue.

There is another school of thought which says that it is not the code, but the team behind the code, that should be considered as the weapon. Farfetched as it is, this again brings the laws of armed conflict back into focus, and the protections afforded to civilians. Civilians have enjoyed the protections afforded by the laws of armed conflict as long as they do not directly participate in the conflict. According to the Tallinn Manual, direct participation includes:

- Conducting cyber attacks, and
- Any actions which make possible specific attacks (for example, identifying vulnerabilities or designing malware specifically to take advantage of particular identified vulnerabilities).

Indirect participation includes:

- Designing malware without the specific intention that it be used in conicts, and
- Maintaining computer equipment generally, even if such equipment is subsequently used in the hostilities.

The Tallinn manual itself has been criticized for a minimalist approach in absolving the State of its responsibilities.

Deterrence in Cyberspace

As an age old practice used to fend off untoward behaviour, deterrence as both theory and practice attracted much attention during the post-World War II era. With the onset of nuclear weapons, a lot was written and debated as new theories evolved explaining the phenomenon to strategic thinkers and decision makers. With the emergence of cyber as a pre-fix to a variety of terminologies—like security, warfare, and even to deterrence—a whole new set of paradigms are being explored, as was the case of nuclear weapons half a century back. Scholars have drawn parallels with nuclear deterrence, finding commonalities, divergences, and convergences between the two.

The cyber dimension has altered strategic thinking in the modern security discipline, as it is one of the key considerations in national strategies for cyberspace or cybersecurity. The ability to deter wide cross-sections of threats is an integral part of such endeavours, which is finding general acceptance as well. Akin to the nuclear aspects of deterrence, there is no dearth of definitions for "cyber deterrence". One of the early works in this sphere has been the monograph *Cyberdeterrence and Cyberwar* by Martin Libicki, which defines cyber deterrence as "...develop[ing] a capability in cyberspace to do unto others what others may want to do unto us."[16] According to Tim Stevens, cyber deterrence "…connotes the use of threats to discourage or dissuade another party from taking actions against oneself, in this context usually understood as a state or, more accurately, the constituent components and assets of such."[17] Cyber deterrence could also be defined as

> a strategy by which a defending state seeks to maintain the *status quo* by signalling its intentions to deter hostile cyber activity by targeting and influencing an adversary's decision making apparatus to avoid engaging in destructive cyber activity for fear of a greater reprisal by the initial aggressor.[18]

In broader terms, cyber means could be employed to deter a hostile activity in cyber or any other domain, by influencing the decision making apparatus of the adversary. Technology and policy means enable deterrence by denial; but at the same time, the cyber domain has opened avenues for nation states to punish the perpetrators of a hostile act whether in the cyber domain itself or in any other domain of warfare, spread across the land, sea, air, and space. Cyber, as pre-fix to deterrence, is a unique challenge to the conceptual understanding and the traditional approaches or theories of deterrence. With peculiar characteristics—whether it is the ambiguity over sovereignty, or the inseparability of civilian and military, amongst others—the very idea of deterrence application in cyberspace has been challenged by the scholars of international relations and war studies.

Critical Infrastructure and Critical Information Infrastructure

From the point of view of a nation state, Critical Infrastructure is a subjective term, encompassing the industries, services, and entities which underpin modern societies and economies. It comprises of the "assets, systems and networks, vital to economic and national security, and public health and safety".[19] Over the last two decades, ever since the US President signed the first directive on Critical Infrastructure Protection, PDD-63, in May 1998, volumes of research in the policy and technology spheres has been dedicated to the subject as a matter of serious concern. Over the same time period, the threat landscape has also changed tremendously. The world has witnessed terror attacks against the World Trade Centre and public transport systems in the busiest metropolitan cities. Sensitive installations, like nuclear facilities and electricity grids, have fallen victim to well-crafted malwares, seen in Iran, Ukraine and elsewhere. Therefore, protecting critical infrastructure is all the while important—in fact, now it has become an integral part of the national security strategy.

The UK government terms Critical Infrastructure as Critical National Infrastructure, and defines it as "those critical elements of infrastructure the loss or compromise of which could result in either a major detrimental impact on the availability, integrity or delivery of essential services or a significant impact on national security, national defence, or the functioning of the state."[20] The elements of infrastructure could be the assets, facilities, systems, networks or processes and even the essential human resources that operate and facilitate them.

The European Union defines Critical Infrastructure as "an asset or system which is essential for the maintenance of vital societal functions."[21] On the similar lines, Australian definition includes "those physical facilities, supply chains, information technologies and communication networks, which if destroyed, degraded or rendered unavailable for an extended period, would significantly impact on the social or economic wellbeing of the nation, or affect the ability to conduct national defence and ensure national security."[22] Essentially, a host of Critical Infrastructure functions are further dependent on information infrastructure for computing, controls, telemetry and communications.

Critical Information Infrastructure, as per a European Union Council Directive, is "the ICT systems that are Critical Infrastructures for themselves or that are essential for the operation of Critical Infrastructures".[23] The IT Act of India defines it as "...the computer resource, the incapacitation and destruction of which, shall have debilitating impact on national security, economy, public health or safety."[24]

The definitions, perspectives, and responses vary from state to state; and so do the definition and scope, all of which are evolving continuously. However, despite variations, the underlying principles, risk to economy, society, or public health and safety, etc. remain, by and large, to be the same. The overall effort of protecting the Critical Information Infrastructure incorporates the prevention of damage, or unauthorised use, or access and exploitation of the information assets underpinning the functioning and operations of Critical Infrastructure.

Critical Information Infrastructure is deemed to be critical because of its positioning and centrality to the functioning of the state. In other words, executing the basic functions of the state is unimaginable without these entities, which are as basic as electricity, water, transportation, telecommunications, and banking services. Moreover, the interdependencies between and among the constituents and cross-sector linkages, etc. make them extremely vulnerable to cascading effects arising out of targeted attacks, failures, or even natural disasters. These distinct attributes or characteristics make them a soft target for hostile actors with malicious intent as the consequences could be farreaching, unpredictable, and possibly catastrophic.

Cyberspace is peculiar in that it dissolves physical and political boundaries, rests upon physical infrastructure but exists in virtual space, and influences the social and cognitive domains. It is ubiquitous, and is often termed as a "man-made" domain, as it forms the fifth domain of warfare. The concepts discussed above are not exhaustive, both in number and scope. In fact, each of the above is attracting vast scholarship and research. Cyberspace does not fall short of the concepts and theoretical frameworks being extrapolated from the disciplines of technology, strategic studies, or international relations.

NOTES

1. "Group of Governmental Experts on Developments in the Field of Information and Telecommunications in the Context of International Security", United Nations, 30 July 2010. Web, 15 August 2014.
2. The National Strategy to Secure Cyberspace, The White House, Washington DC, February 2003, p. vii.
3. Ronald R. Fogleman, "Information Operations: The Fifth Dimension of Warfare," Remarks delivered by Air Force Chief of Staff to the Armed Forces Communications-Electronics Association, USA, Washington DC, 25 April 1995, Web, 12 November 2015, at http://www.iwar.org.uk/iwar/resources/5th-dimension/iw.htm.
4. "The Definition of Cyberspace", Deputy Secretary of Defense Memorandum, 12 May 2008.
5. Rauscher, Karl Frederick, and Valery Yashchenko, "Russia-US Bilateral on Cyber Security: Critical Terminology Foundations", New York and Moscow: EastWest Institute and Moscow State University, 2011, p.16.
6. Ibid.
7. "India: National Cyber Security Policy 2013", Department of Electronics and Information Technology, Delhi: n.p., 2013, Web, 14 November 2015.
8. "Information Technology—Security Techniques—Guidelines for Cybersecurity", International Standards Organisation, 2012, Web, 14 November 2015, at https://www.iso.org/obp/ui/#iso:std:iso-iec:27032:ed-1:v1:en>.
9. Ibid.
10. "Explore Terms: A Glossary of Common Cybersecurity Terminology", National Initiative for Cybersecurity Careers and Studies, US Department of Homeland Security, at http://niccs.us-cert.gov/glossary.

11. "New America Global Cyber Definitions Database", at www.newamerica.org/cyber-global/cyber-definitions/.
12. Richard A. Clarke and Robert Knake, *Cyber War*, Harper Collins, 2010, p. 228.
13. U.N. Charter, art. 2, para. 4
14. Matthew Hoisington, "Cyberwarfare and the Use of Force Giving Rise to the Right of Self-Defense", *Boston College International and Comparative Law Review* 32, 2009, p. 447.
15. "Definition of Aggression", United Nations, December 1974, Web, 23 November 2015, at http://www.un-documents.net/a29r3314.htm.
16. Martin C. Libicki, "Cyberdeterrence and Cyberwar", RAND Corporation, 2009, at https://www.rand.org/content/dam/rand/pubs/monographs/2009/RAND_MG877.pdf, p. 27.
17. Tim Stevens, "A Cyberwar of Ideas? Deterrence and Norms in Cyberspace", *Contemporary Security Policy*, Vol. 33, No. 1, 2012, pp. 148–170, p. 151.
18. Emilio Iasiello, "Is Cyber Deterrence an Illusory Course of Action?" Journal of Strategic Security, Vol. 7. No. 1, 2014, pp. 54–67.
19. "What is Critical Infrastructure?" US Department of Homeland Security, at https://www.dhs.gov/what-critical-infrastructure, accessed 02 May 2018.
20. "Critical National Infrastructure", Centre for Protection of National Infrastructure, at https://www.cpni.gov.uk/critical-national-infrastructure-0.
21. "Critical Infrastructure", European Commission, at https://ec.europa.eu/home-affairs/what-we-do/policies/crisis-and-terrorism/critical-infrastructure_en.
22. "Critical Infrastructure", Trusted Information Sharing Network, at https://www.tisn.gov.au/Pages/Critical_infrastructure.aspx.
23. "Critical Information Infrastructures", European Union Agency for Network and Information Security, at https://www.enisa.europa.eu/topics/critical-information-infrastructures-and-services/cii.
24. "The Information Technology (Amendment) Act, 2008", Ministry of Law and Justice, http://www.meity.gov.in/writereaddata/files/itact2000/it_amendment_act2008.pdf, p. 13.

CHAPTER 2

Cyber Deterrence: The Emerging Landscape

As cyberspace has become more integrated into virtually every facet of human existence, assets in cyberspace have become attractive as potential targets against which attacks can be carried out to weaken other states. The current attributes of cyberspace are such that cyber attacks can be carried out with very little chance of retribution, the main reason being that it is very difficult to draw a definitive conclusion on the geographic source of such attacks. The end result has been increasing instability in cyberspace, the reduction in its utility, and the weakening of confidence in users of its resilience. Being an integral part of statecraft and domestic as well as foreign policy making or military strategy, there have been many attempts by various scholars to apply deterrence and its associated concepts to cyberspace.

In the modern history of statecraft and military planning, the concept of deterrence evolved with nuclear weapons, theories, analyses, mathematical models which gave precise projections of the expected fallout from the use of nuclear weapons as well as numerous debates shaped strategic thinking on the subject, and its impact on international relations. Out of these intense debates and strategic competition, stemmed the concepts of counterforce (military) and counter value (civilian, major infrastructure, etc.), "credible minimum deterrence" and the modicum of "deterrence stability".[1]

Given the success of deterrence in reducing the chances of a nuclear conflict significantly and, in the absence of a similarly viable concept to regulate conflict in cyberspace, there have been many attempts to adapt deterrence concepts to cyberspace. However, deterrence does not easily adapt itself to the domain of cyberspace and state conflicts.

While there are some similarities between nuclear conflicts and cyber conflicts—such as the potential for large-scale effects—there are features so unique to each of the nuclear and cyber arenas which makes comparing them as difficult as comparing apples and oranges. This makes deterrence a complex concept for cyberspace even if it is modified to suit its unique features. For deterrence to be credible, threats of severe retaliation require attribution and a quick response, both of which are plausible in the nuclear arena. However, at present, attribution to the extent necessary in a deterrence matrix is not possible for cyberspace. Other factors inhibiting the use of the deterrence concept in cyberspace include the proliferation of actors with different risk appetites, and the fact that cyber weapons are very different compared to nuclear weapons, which can be precisely quantified in terms of tonnage as well as in terms of the physical damage or adverse effects they can cause.[2]

Apart from the fact that there is no clear definition of what constitutes a cyber weapon, the development of such weapons is shrouded in secrecy; so also is their deployment. This is because, unlike nuclear weapons, they are not in physical form. Again, unlike in the nuclear arena, where the availability of the weapon is known or can be computed—and this in itself acts as a deterrent—the presence of cyber capabilities does not seem to act as a deterrent as evidenced by the continued cyber attacks against the USA.

Deterrence requires foreknowledge about the possible perpetrators. However, the nature of cyberspace is such that there is usually very little clarity about attackers even days or months after a sophisticated attack, where the identity and location of the source is deliberately concealed. What further complicates the situation is that the source of an attack could be a state, organisations, or individuals working on behalf of a state, or even independently. The personnel could have criminal or activist antecedents. They could also be operating from another state, or even from within the targeted state itself. Existing international law has limited options to deal with such situations. All these variables make it difficult for deterrence which is based on concrete presumptions.

Analysts and experts are coming around to the opinion that while deterrence is a concept worth exploring in the absence of alternatives, the concept of deterrence as applied to cyberspace needs modification. In the first instance, more emphasis has to be placed on the various types of actors in cyberspace, and the *de-facto* deterrent power that they possess since states find themselves powerless to act against them. New formulations have to be developed, taking into account the various types of actors and the various types of threats, as well as identify the best ways to deter the respective actors and neutralise various threats. According to Joseph Nye, deterrence has to envelop a broader spectrum of activities ranging from deterrence by punishment, deterrence by denial, entanglements, and normative taboos to make it viable.[3]

As India embarks on the path of digitalisation and harnessing cyberspace for its legitimate interests, it has become imperative to achieve the ability to deter cyber attacks. At present, India has only limited capabilities in the spectrum of activities outlined by Joseph Nye and other scholars to deter attacks in cyberspace.

Principles of Deterrence

Deterrence has been an age old practice in statecraft and military strategy. It was exercised to discourage any sort of untoward or unwanted behaviour in an adversary to prevent conflict. Earlier, strong states used to discourage their potential adversaries from attacking them, not just with the threat of defeat but by raising the cost of the conflict to such a level that attacking becomes an unacceptable option for the adversary.[4] Punishment, therefore, enabled deterrence. With the onset of nuclear weapons, the role of punishment in deterrence was further enhanced; however, it was not created anew. Prior to nuclear weapons, naval or terrestrial blockades and aerial bombing raids[5] were the plausible options to inflict punishment on the adversary for an untoward action. However, the concepts, theories, and its practice as an instrument of foreign and security policy emerged and gained traction with nuclear weapons.[6]

As a subject of analysis and intense debate in the discipline of International Relations, deterrence has attracted vast scholarship and academic literature after World War II. Thereupon, deterrence has evolved into an elaborate strategy, and has become a distinct way in itself of pursuing national security.[7] As deterrence was practiced throughout the Cold War, comprehensive academic research and frameworks were developed in the specific context of nuclear deterrence.

Deterrence, in essence, tends to change or alter the decision-making calculus of the adversary, either by increasing the perceived costs of their action (deterrence by punishment) or by decreasing the expected benefit (deterrence by denial).[8] In the words of Glenn Snyder, deterrence means "discouraging the enemy from taking military action by posing for him a prospect of cost and risk outweighing his prospective gain."[9] Deterrence and defence are inherently different, where deterrence works on the enemy's intentions, which is fundamentally a peacetime objective to preserve the *status quo*; defence, on the other hand, is carried out to reduce one's own prospective costs and risks in case deterrence fails, and therefore, holds a wartime value.[10] Patrick Morgan defines deterrence as the "use of threats to manipulate behaviour so that something unwanted does not occur."[11] The very conception of deterrence is to prevent an attack by threatening the adversary with damage to the extent that the best choice in the adversaries' cost-benefit calculus is not to attack.[12]

Deterrence, as Lawrence Freedman has put it, "is to persuade the enemy through denial that the envisaged gains would be hard to come by, and whatever gains might be obtained would soon be outweighed through punishment."[13] While taking a cognitive approach to deterrence, Robert Jervis defines it as "a psychological phenomenon, which involves convincing an opponent not to attack by threatening it with harm in retaliation."[14] Being a psychological phenomenon, deterrence succeeds or fails in the mind of the potential attacker. As per Jervis, it is the persuasiveness of the message *per se* about the retaliatory capabilities of the nation state rather than the capabilities themselves that determines the success or failure of deterrence.

Tracing the use of the term 'deterrence' in Russian discourse, Dmitry Adamsky places it as compellence, which includes the general prevention of the threat from materialising, therefore deterrence in peacetime and the use of force during wartime to shape the battlefield. The Russian understanding of deterrence is not as a brute force strategy but as coercion aimed at manipulating the perception of the adversary and influencing its strategic behaviour[15]—it is more of a psychological phenomenon. By and large, the basic tenets of deterrence are common; the perspectives originating from different countries might vary.

Deterrence, as a concept, also rests on assumptions, particularly about the way potential adversaries recognize, interpret, and react to the threat(s) of retaliation. Correct recognition and interpretation of threat of retaliation is essential for the adversary to gauge the risks and prepare the course of action appropriately. If the potential adversary does not recognize the threat of retaliation in the first place, it is not likely to be deterred from taking the risky course of action. Adversaries are also prone to either underestimate or overestimate retaliatory capabilities. The former renders retaliation threats futile while the latter runs the risk of escalation in a conflict.

Denial and punishment are the two prime underlying principles of deterrence. Denial mechanisms intend to convince potential attackers or adversaries that their chances of succeeding in an attack are feeble. The deterring state has to ensure that the effort and cost required to launch a successful attack against it outweighs the anticipated benefits. Deterrence by punishment ensures the fear of a strong response to an attack, and the scale or severity of the retaliation might inflict more harm than the adversary is actually willing to bear. The doctrine of mutually assured destruction during the Cold War is one of the apt examples of deterrence by punishment. The mere threat of using a nuclear weapon against themselves prevented the adversary from using a similar weapon. Also, in the words of Lawrence Freedman, when the possibilities for building and executing denial are limited and vulnerability is inevitable, deterrence has to work through punishment.[16]

Even the concept of the balance of power was based on deterrence, carried out through wars as well as to prevent wars.[17] Patrick Morgan makes an enticing observation that deterrence has ever since been employed as part of security strategy. During the Cold War—and even after that—extending deterrence became central to international politics, involving alliances, interventions, arms transfers, power projection efforts, military training programmes, and non-proliferation pressures. Even collective actor deterrence is an extension of the emerging version of international law and order, and the continuing development of norms—transnational, international, and domestic—of acceptable state behaviour.

The way international efforts and opinion on cybersecurity are shaping up towards norms development, an extension of deterrence is somewhat taking the shape of collective actor deterrence. Even Tim Stevens has made the observation that, in the international policy sphere, US led initiatives have focused primarily on the development of normative frameworks for behaviour in cyberspace. Deterrence, as theory and conceptual model in international relations, finds its underpinnings in the Cold War era, where nuclear weapons dominated the security calculus among the two major powers of the world. The discussions and debates around cyber deterrence have gained traction in the geopolitical context, specifically in the aftermath of alleged Russian interference in the presidential electoral process of the USA, and other massive attacks on governmental entities like the Office of Personnel Management or the health sector. Extensive scholarship on cyber deterrence over the last one decade, since the attacks on Estonia in 2007, has graduated to conceptualizing cyber deterrence, while states like the USA are practicing deterrence in cyberspace with a cyber command, through the North Atlantic Treaty Organisation (NATO) and also through the international norms building exercises.

As a matter of fact, much of the conceptualization and discourse on deterrence derives from the West. But, the concept of deterrence is also deeply rooted in strategic and military thinking across Russia, China, and India. In a fundamental departure from the prevalent thought on deterrence in general, Russian strategic thinking is cross-domain, as it aims not just the prevention of aggression but also to influence the behaviour of the adversary in other fields of activity. Dmitry Adamsky observes that Russia has de-emphasised the nuclear aspects of deterrence, and moved towards 'non-nuclear deterrence', underpinned by the belief that non-nuclear means precision weapons, ballistic and cruise missiles, and informational or cyber capabilities that can generate deterrence effects compatible with nuclear weapons.[18]

In Chinese, *wēishè* is the term which resembles the meaning of deterrence.[19] However, as per Dean Cheng, *wēishè* embodies both dissuasion and coercion and, in spirit, it is viewed as a means to achieve political ends. Based on Chinese writings, Dean Cheng concludes that the Chinese understanding of deterrence is also all-pervasive and, in practice, it may be incorporated through conventional, nuclear, or cyber/information domains, augmented by using economic, diplomatic, and legal channels.[20]

The discourse on deterrence in India, took shape in the backdrop of the 1998 nuclear tests, after which India developed its nuclear doctrine. In modern history, the writings, strategic analyses, and publications on the subject of deterrence have had a nuclear influence. Even the corresponding doctrinal or strategic developments have been under the shadow of nuclear weapons. Two decades later, cyber has emerged as a domain with a visible impact on military and strategic thinking. The approach to deterrence has been shaped by national policies, perspectives, and experiences, but cyber has added a new context, compelling nation states to either bring in doctrinal changes or meld it with their traditional practices of exercising deterrence.

Conceptualizing Cyber Deterrence

Professor James Der Derian had coined the term "cyber deterrence" in a 1994 issue of the *Wired* magazine, contextualizing the presumed deterrent effect of network and information technologies on the physical battlefield.[21] Around two and a half decades later, cyber deterrence has emerged not only as a subject of academic discourse, but rather as a full-fledged security discipline. Joseph Nye has written extensively on the concept of deterrence and its applicability to the cyber domain, arguing that mitigating the risks from myriads of threats is a complex mechanism, encapsulating threats of punishment, denial, entanglement, and norms of behaviour.[22]

Will Goodman has analysed the tenets of cyber deterrence under the framework of eight elements of deterrence, namely: an interest, a deterrent declaration, denial measures, penalty measures, credibility, reassurance, fear, and a cost-benefit calculation.[23] If one considers the case of a nation state seeking to deter an adversary, it is basically protecting an interest, which could be political, economic, social, or any other thing valuable to the state. In the quest to keep the adversary away, the state issues a deterrent declaration. At the same time, the state employs both denial and penalty measures to prepare a response in case deterrence fails. Also, the deterrent declaration has to be credible and reassuring so that the adversary considers it potent. The penalties must instil fear of loss and damage, and the responses should tilt towards serious unacceptable costs in the cost-benefit calculation of the adversary.

By way of comparison, cyber deterrence, *per se*, does not find mention in Russian literature; but it is seemingly encapsulated in the concept of "strategic deterrence". In 2008, General Makhmut Gareev, the President of the Academy of Military Science, introduced the concept of strategic deterrence encompassing interrelated political, diplomatic, information, economic, military, and other measures that deter, reduce, or avert threats and aggressive actions.[24] S.G. Chekinov and S.A. Bogdanov have also written on the concept of Strategic Deterrence in Russia. In a publication in *Military Thought*, a journal of the Ministry of Defence of the Russian Federation, they term strategic deterrence as a unified concept encapsulating measures in the domains of politics, economics, ideology, information, technological science, military and non-military steps, which may be adopted consecutively or simultaneously to dissuade the adversary's military and political leaders and the public, from achieving their military and political goals by force.[25] Strategic deterrence, therefore, works towards stabilizing the military and political situation, exercised at the global and regional levels.

Invoking the imperatives of the human element in the entire process of deterrence, Dorothy Denning deems deterrence as fundamentally influencing the decisions and actions (or inactions) of decision makers who are actually human beings.[26] She has aptly underscored a few reasons why the concept of cyber deterrence is extremely challenging to both conceptualize and to put into practice or action. The term itself, as per Denning, is extremely broad, and encapsulates the entire domain of warfare as there is no notion of "land deterrence," "sea deterrence," "air deterrence," or "space deterrence". Rather, deterrence is focused at a particular weapon or a specific activity—like nuclear deterrence is basically about a specific type of weapon and the means to deliver, it cutting across all domains of warfare (land, sea or air), but is not a domain in itself.

This situation is somewhat akin to the one from more than half a century ago when nuclear weapons first came into existence. The initial emphasis was on acquiring the requisite technologies and stockpiling nuclear weapons. However, building up a fool proof defence against nuclear weapons proved to be an impossible task, particularly since only one had to get through even the most impregnable of shields to cause destruction on a mass scale. This led to the concept of deterrence where the emphasis was on conveying to adversaries the capabilities to inflict damage in the event of a provocation. Vast scholarship has drawn analogies between nuclear and cyber deterrence; however, the application of concepts, theories, and practise of nuclear deterrence are not deemed fit for cyber deterrence, given the differing context, properties, and characteristics.[27] A stark difference between nuclear and cyber deterrence is that the former works because the number of countries in possession of such weapons are restricted due to the significant resources and the technological know-how required to raise them. Even those who possess nuclear weapons understand and recognise the potential of such weapons to cause widespread destruction. In addition, there are international arrangements, regimes and institutions (the Treaty on the Non-Proliferation of Nuclear Weapons, the

Nuclear Suppliers Group, and the International Atomic Energy Agency) to counter proliferation and the risks from the use of nuclear weapons. There are serious barriers of entry even for nation states; non-state actors stand only a remote chance. As opposed to deterrence by denial, (which depended on fool proof defences against the delivery mechanisms of nuclear weapons, be it missiles or aircraft), deterrence by punishment held the peace through much of the Cold War.

In the cyber domain, countering proliferation is arduous, both due to technical as well as institutional reasons. Software programs can easily be sent, shared, transferred, or even replicated over computer networks, and between all sorts of actors, be they individuals, criminal syndicates, or even nation states.[28] Technical measures to curb proliferation are ineffective and inadequate, and there is an absolute dearth of international arrangements or institutions to do so. Another dimension which makes nuclear deterrence effective, and renders cyber deterrence inept in comparison, is the magnitude of the destruction nuclear weapons could inflict upon the states. The perils, risks, and consequences are well understood by the decision makers. However, these are not comprehended or foreseen in the case of cyber based attacks, which are perceived to remain below catastrophe levels as seen in the context of disruption and monetary loss.

Tim Stevens holds a strong view that, given peculiar characteristics and conditions, the procedures and techniques of Cold War deterrence[29] cannot be extrapolated to the cyber domain. There are also numerous operational challenges and functional ambiguities around cyber deterrence. It is not yet as mature as nuclear deterrence, in addition to the fundamental differences between both. The radical change nuclear weapons brought to national security strategy is not just because of the mass or scale of destruction, but also the reduced time for response or retaliation. With blockades and bombings, punishment to an unwanted action was slow, leaving plenty of time for bargaining and negotiations. Escalation under those settings had less probability. Responses in the cyber domain, similar to the nuclear option, could be swift and devastating, raising the risks of escalation within the domain, or even across the domains.

In the case of cyberspace, the tilt of deterrence is more toward deterrence by denial. This essentially calls for the deployment of robust defensive measures to discourage an adversary from attacking. Cyber defences are also becoming proactive, involving the extensive use of counter espionage, or intelligence extending to the use of pre-emptive operations. With elevated defensive or denial mechanisms, retaliatory options are also increasing in leaps and bounds, and are certainly not just restricted to the domain of cyberspace itself. Joseph Nye argues that deterrent threats need not be limited to responses in cyberspace; they could well be extended to other domains to dissuade both the general behaviour of aggression as well as some specific acts of the proliferating threat actors.

Threat Actors in Cyberspace

A deterrence strategy for cyberspace has to address four broader sets of threats, emanating from terrorism, crime, espionage, and asymmetric attacks targeted at the critical infrastructure.[30] The actors behind these threats have different capabilities to impose harm, and varying degrees of tolerance for risk to their own operations or infrastructure. For instance, a nation state is more prone to risks from retaliatory attacks on its own critical infrastructure which could endanger its populace, while a terror group is immune to those risks as it does not hold territory, own infrastructure, or have a population to defend against retaliatory attacks. In retrospect, non-state actors are the most difficult adversaries in cyberspace to deter, as they do not have territory, population, or political constraints, which are extremely valuable for nation states, and also happen to be the key determinants of a deterrence strategy.

However, both Joseph Nye and James Lewis do not deem cyber threats to be existential threats for a nation state,[31] assuming territorial integrity and political independence as the vital interests of a nation state.[32] James Lewis argues that cyber attacks that do not pose existential threats are hard to deter—like cyber espionage or cyber crime—as they fall well below the threshold to be justified for a military action in response.[33] Off late, particularly in the wake of alleged Russian interference in the Presidential elections of the USA,[34] cyber has emerged as a preferred domain for meddling in the political processes, specifically electoral processes, of other nation states. Cyber means or social media to influence public opinion, either to benefit or undermine the political interests of the candidates in democratic political environments certainly pose a serious and potent threat.

In order to be effective, cyber deterrence must be targeted at the entire spectrum of threat actors, varying from individuals, criminal syndicates or organisations, to nation states. As Eric Talbot Jensen argues, the state should consider the full spectrum of actions to attain deterrence against cyber threats, and these vary from small invasions or disruptions to large scale cyber attacks; however, it must be proportional to the threat or use of force anticipated.[35] In cyberspace, all threat actors are quite capable of inflicting significant damage, and the spectrum of potential adversaries is wide. The responses as per a deterrence strategy need to apply to the full spectrum of threat actors and the entire spectrum of actions, from small invasions or disruptions to the possibilities of large scale attacks in cyber or other domains.

In cyberspace, state actors pose the most credible and potent threats, with their militaries, intelligence agencies, and other secret security services for political, military, economic, or industrial espionage, even extending to acts of disruption and destruction. State led espionage operations do not just target information about the governments, their armed forces, or diplomatic engagements; they extend, in certain cases, to the theft of intellectual property, technology, trade secrets, or advanced research programs. Nation states have the requisite wherewithal, resources, and expertise to engage their adversaries in the cyber domain.

However, non-state actors may probably lack the resources or infrastructure; but their technical expertise is to be reckoned with. Although non-state actors usually comprise of terrorist organizations, criminal enterprises, or individuals, they are increasingly playing an important part in augmenting the capabilities of nation states as they are being employed by state agencies for specific tasks. This also enables plausible deniability for nation states at the time of the attribution of cyber attacks, and brings in the hybrid aspects. The collusion between nation states and the non-state actors fills in the capability and skill set gap for nation states. Non-state actors sometimes also act as proxies for the state.[36] Alliances are quite possible among non-state actors to complement the capabilities of each other, and could also act as force multipliers[37] with ease in scaling and access to vast pools of human resources to achieve common goals, or even for monetary gains. Different threat actors in cyberspace have different vulnerabilities of their own, different strategies and modus operandi, and different risk exposure. Deterrence, as a strategy in both general and specific terms, has to cover this wide spectrum of threat actors amidst the changing geopolitical environment.

The Geopolitical Perspective of Cyber Deterrence

Drawing cues from the analyses of the cyber attacks in the past, it is quite plausible that they will be employed in the context of any ongoing geopolitical tension or conflict. This was seen in the case of Stuxnet, the Distributed Denial of Attacks on Estonia in 2007, the Georgia War in 2008, the hacking incident at Sony Pictures in 2014, and at the Ukrainian electricity grid in 2015, to name a few. All these instances emerged in the backdrop of an ongoing geopolitical stand-off or conflict. Stuxnet took off during the conflict between the USA and Iran over its nuclear weapons development programme. The alleged three-week long Russian DDoS attacks on the Estonian parliament, banks, and broadcasters were in the backdrop of the decision to move a memorial to the Soviet Red Army at Tallinn, to a new place. The USA-China diplomatic standoff over the indictment of Chinese military hackers on the charges of cyber espionage was a result of the massive scale of economic espionage against US intellectual property. Similarly, the prominent attacks over the last one decade were in response to an open geopolitical contest or conflict, opened as a new front to act as a force multiplier. Interpretation of the geopolitical context requires connecting the dots from the extensive knowledge of historical, political, technical, economic and military aspects of the threat actors.

On the one hand, cyber attacks have geopolitical characteristics; on the other hand, to be effective, cyber deterrence requires geopolitical attributes. Will Goodman argues in favour of geopolitical symmetry among the states for cyber deterrence to work effectively. It is pertinent for the state to protect itself, if and when the conflict in cyber domain escalates beyond its domain, and spills over to the physical domain.[38] Taking cues from the case of Estonia, Goodman elaborates that asymmetry between Estonia and Russia, in physical terms, limited the responses of Estonia during the 2007 cyber attacks. The sheer inclination of asymmetry in favour of Russia deterred any sort of Estonian retaliation. Later, however, NATO's cyber defence policy of 2008 and the developments thereof[39] were an Estonian bid to bring equilibrium to geopolitical disparity vis-à-vis Russia, with NATO involvement in a combined cyber defence.

The international community has taken some steps to address the challenges arising out of the cyber domain. The Budapest Convention is one of the first steps taken in this regard to tackle cyber crime. Cybersecurity is on the agenda for most multilateral discussion forums, be it the Association of Southeast Asian Nations (ASEAN), the ASEAN Regional Forum (ARF), the Asia-Pacific Economic Cooperation Organization (APEC), or the Organization for Economic Co-operation and Development (OECD).[40] With cyber defence being a part of NATO's political agenda, collective security arrangements and the coordination of armed forces have also picked pace over the last one decade.

The next section of this chapter explores the overarching issues which are being discussed across the globe in the context of developing the concept, framework, or strategies related to cyber deterrence.

Making Deterrence Functional: Key Challenges for Policy Making and Overarching Issues

Every technology enabled nation is dependent on cyberspace, from the delivery of essential governance services, or facilities in banking, telecom, healthcare, etc. to conducting business operations in a globalised world. Attaining the ability to defend one's own networks and, at the same time, dissuade any adversary from gaining unfair advantages through acts of crime, war, or espionage is what the entire policy making and law enforcement apparatus is striving for. The issues of attribution, proportionality, and communication (among others) are prominent, and are increasingly being deliberated in academic and strategic circles.

As a part of the national security strategy, the aim of deterrence is to create disincentives for starting or carrying out a hostile action. Arguing for tailored deterrence in addition to cyber defence, Franklin D. Kramer and Melanie J. Teplinsky have discussed a hybrid model of cybersecurity. Such a model lays heavy emphasis on raising the costs of, and reducing the benefits from, cyber attacks.[41] At the bare minimum, the endeavour is to secure and safeguard one's own networks and assets—which is more of a defensive measure. Heightened security keeps a major proportion of threat actors at bay, and ensures that the critical assets, information, data, and equipment is out of the reach of attackers. Even if an attacker attempts to intrude or perpetrate an attack, the defences ensure that the cost, time, and effort become quite significant for the attacker so that the option of aggression becomes unviable for the attacker. The threat of retaliation also dissuades the potential adversary to initiate any act of aggression which the defender does not want the adversary to take. Deterrence and defence are fundamentally different, and so are their strategies. Defensive policies or strategies are designed to fend off an adversary in the event of an attack, while deterrence intends to convince an adversary not to attack in the first place.[42]

Deterrence by Denial

Deterrence by denial employs various means to strengthen defences in such a manner that the efforts, resources, and costs required for a successful attack are enormous. There are different methods and approaches used to do this. Some are discussed below:[43]

- **Enhanced Cybersecurity**: Enhanced cybersecurity is more like a security ring which fends off a majority of the attacks before they can achieve their goals. The approach to cybersecurity generally includes stringent authentication and password management, encryption of data and communication channels, analysis and assessment of viruses or malware, and the timely update or patching of software for known vulnerabilities.
- **Active Cyber Defence**: Active defences in the form of network monitoring or surveillance for the swift identification of and counter measures against cyber attacks are gaining prominence since the ways and means used are typically moving beyond traditional cybersecurity practices. These include monitoring network traffic, blocking hostile packets, and deploying honeypots. Active defence for network security also helps in unveiling the identity of the perpetrators of an attack, as well as facilitating justice and prosecution in accordance with the respective legal frameworks.
- **Redundancy and Resiliency**: Redundancy in infrastructure ensures the sustainability of operations in the case of attacks or other disasters/ accidents which degrade infrastructure. Redundant assets remain functional under such contingencies, containing the propagation of failure or disruption. Although building redundancy and resilience into network systems adds to the costs and architectural complexity, they are quite effective in mitigating operational risks to a larger extent. Well defended and resilient information systems and computer networks can reduce the perceived gains from a cyber attack for the adversary. Enhanced defence mechanisms could be further reinforced or supplemented by multilateral arrangements for acceptable behaviour or norms in cyberspace.

- **International Norms of State Behaviour for Cyberspace, Conflict Prevention, and Confidence Building**: Diplomatic measures to prevent conflict and build confidence among the stakeholders in cyberspace are a cornerstone of stability in this domain. Such activities are actively being pursued at global and regional levels such as in the United Nations, the International Telecommunication Union, the ASEAN Regional Forum (ARF), and OECD, etc. with focus on practical measures to build confidence among member states or pave the way for norms of behaviour in cyberspace. Norms are instrumental in pressurising or adding costs on the attacker, and reducing the so-called merits and perceived benefits from the attacks.
  Experts have consistently argued in the favour of international norms to underpin cyber deterrence. Catherine Lotrionte supports multilateral arrangements on acceptable behaviour in cyberspace, during both peace and war time, to augment the cyber defences of nation states. The constraints arising out of norms affect political decisions related to risk assessment and the costs or benefits from cyber attacks.[44] Using the example of Nuclear Non-Proliferation Treaty of 1968 for international norms and agreements to limit the acquisition and use of nuclear technologies, Dorothy Denning underscores the cumulative effect of denial as well as norms on deterrence dissuading states from even attempting to acquire nuclear weapons.[45] However, diverging interests, varying cultures of norms and behaviour, in addition to the practical challenges of verification make treaties extremely difficult to negotiate and enforce. Simultaneously, deterrence poses challenging questions before the scholars of international relations and international law in terms of understanding the possibilities of norms becoming legally binding international laws as well as the modalities surrounding compliance with such laws.
- **Entanglements**: Economic, political, social, or other spheres of interactions and engagements lead to entanglements. These interwoven dependencies make the attacker question the very necessity or attractiveness of the attack as it may result in severe damage for the attacker himself. Entanglements mould the attacker's perception of the targeted system as emanating interdependencies might significantly impact its own infrastructure or assets which the attacker values.

Joseph Nye argues that entanglement in symmetrically interdependent relationships can still lead to deterrence even in the case of inadequate attribution because a major attack would be counterproductive for the attacking state itself. As per Nye, economic interdependence as well as social, scientific and diplomatic interactions, etc. has entangled countries in multiple networks—the USA and China are the prime example. A massive attack harming the economic interests of the USA would, in turn, also harm China, and viceversa.[46] Possibly, entanglements lead to a situation akin to mutually assured destruction in the economic sphere, and thus acting as a deterrent. However, this would only work for roughly symmetric powers that have equal amounts to lose from such an economic confrontation. This could also form the basis of extended deterrence since if a country had substantial investments in another country, it would be in its interest to ensure the economic wellbeing of that country.

Deterrence by denial has multiple modes of execution in the various spheres of technology, diplomacy, policy, and strategy. Denial mechanisms are an outcome of a blend of these measures. Nevertheless, if defensive measures fail to contain an adversary from perpetrating an attack, the threat of punishment warrants the adversary to refrain from taking any untoward action. Strong defences make an attack exceedingly difficult for the attacker; but the threat of severe retribution discourages prospective or potential attackers.

Deterrence by Punishment

In order to elevate the level and the credibility of deterrence, despite heightened defences, the defender needs to threaten the wide spectrum of malicious actors with punishment for any unwanted or undesirable behaviour. At a minimum, deterrence requires the ability to distinguish between good and bad behaviour confidently. False positives and false negatives are both detrimental for such a policy. Undeserved punishment lacks legitimacy, and the failure to punish guilty appropriately weakens the deterrence posture.[47] Incorrect attribution of an incident runs the risk of directing the responses towards the wrong target.[48] Deterrence by punishment has many overarching issues which lie at the crossroads of legalities and technicalities of the options available to a nation state.

Attribution

Computer networks, on which the very basic functioning of Internet and associates services rests, were designed for an open and trust based methodology for the further development of this technology. As it grew thereafter in the form of cyberspace, it enabled anonymity for the users to a greater extent, thus becoming one of the key pillars on which the Internet rests today. Anonymity is the ability to mask one's identity on the internet by using multitude of obfuscation techniques, such as anonymizers and proxy servers, Onion Routing techniques, or exploiting the loopholes in network architecture. Despite being of great value for governments, businesses, and the individuals, the Internet is being used aggressively for malicious activities. This undermines not just the security of the state, the individual, and institutions, but also the ingrained trust in the Internet as a technology and service platform. The primary roadblock for the attribution of cyber attacks stems from the inability to verify the presence of the alleged perpetrators and/or their sponsorship thereof. This is because masking the tracks always leaves space for plausible deniability. Despite repetitive allegations against China for economic espionage against high value targets in the USA, the Chinese government has always denied its indulgence in any such activity.

David Wheeler defines attribution as "determining the identity or location of an attacker or an attacker's intermediary."[49] Attribution with high probability is core to the practice of deterrence,[50] and cyber deterrence is no exception. In fact, in the context of cyber deterrence, foremost attention is always drawn to the problem of attribution. Despite practical challenges and ambiguities surrounding attribution, the efforts to do so—either technical or political—remain vital for cyber deterrence. Correct and timely attribution brings legitimacy to retaliatory action and enhances credibility; both are quintessential to justify responses on the domestic and international fronts. Attribution is the central component of deterrence strategy,[51] and it is necessary to be established before the victim or defender commences any retaliatory action through diplomatic, economic, or military options in the case of nation states.

Moving a step further, Clement Guitton argues in his recent book that attribution is not merely about the identification of attackers, but is the whole process of unravelling the entire chain of individuals, organizations, and also states involved in the attacks.[52] This calls for a multidisciplinary analysis of technical, political, cognitive, and behavioural factors, in addition to the geopolitical context of the attacks. Technical solutions fall short in correlating the interaction between the actual attackers and the mandating state behind the attacks. Technicalities of digital forensics need to be analysed in consonance with the existing geopolitical context, reinforced with non-technical means of intelligence gathering.[53] Also, attribution is not a binary of "solved" or "unsolved"; it is a process, is measured in varying degrees,[54] and spans the triad of technical, legal, and political dimensions.

Attribution is not restricted or contained to a few experts or institutions, especially with the rising complexity of malwares used for cyber attacks. Even malware analysis, as the beginning step, requires diverse skill-sets to analyse specific components like command and control, payload, or the propagation. Attribution has a wider ambit and different levels as it encapsulates digital forensic investigators, the intelligence community, the executive arm of the government, political leadership, journalists, industry experts, and scholars or research analysts from the academia. It is collective effort and expertise which enables effective attribution. At the operational level, expertise lies in forensics, industry, and academia, while strategic analysts, the political leadership, and the executive from foreign affairs, national security apparatus, or the defence establishment form the strategic layer. Also, international cooperation is vital for attribution, particularly given the transnational nature of cyber attacks. Enhanced cooperation and interaction among these individuals, institutions, and establishments is deemed critical to establish attribution. Attribution works both ways; technical analysis, forensics, and evidence feed into the governmental apparatus for strategic calculations, and conversely strategic imperatives guide and drive technical analysis, digital evidence collection, and forensics.

Expectations and deliverables are different at various levels. For instance, the tactical level is concerned with the technical analysis of the incident, its components, damage assessment, and the execution itself. The operational level deals with the architecture as well as the attacker's characteristics and attributes to figure out what the attacker has been looking for.[55] Analysis at the strategic level disentangles the perpetrators, their motives, and the rationale of the attack. This primarily answers the questions of who could possibly be the instigator, and why he has launched the attack.

As the capabilities and technologies with the intelligence or law enforcement agencies or the governments *per se* are graduating, attribution is further strengthened technically. The private sector is also complementing this process, and a niche market has emerged out of it with improvised services and products. Private entities are playing an increasingly important role, in both technical and diplomatic spheres. Mandiant is one of the prime examples how private sector entities can play an important role in this burgeoning arena. Over the last 6-7 years, Mandiant has analysed Chinese espionage networks and worked closely with the US intelligence community. It has not just been a contributor to the technical investigations but also, in a way, shaped the agenda of the diplomatic engagement between the USA and China on the overarching issue of cyber espionage. Attribution is pertinent for retaliatory measures, be they political, diplomatic or even military. However, it is not a significant factor in the others means of enabling cyber deterrence, be it denial, entanglement, or normative frameworks.

Scope and Methods of Retaliation

The larger question before decision makers pertaining to retaliation in any strategy for cyber deterrence is the decision to use military force in response to a cyber attack. The question of classifying a cyber incident to be "an act of war" or as a "use of force" under the existing international law has not found any answers yet. A military response is justified for an incident adjudged to be an act of war.[56] In this context, there is no dearth of techniques with adversaries to reduce the risks of retaliation by employing proxies,[57] or to restrict their cyber attacks under the threshold that would qualify it as an act of war or the use of force. This restricts retaliatory options and seriously undermines the deterrent capability. Restricted scope and options erode the deterrence effect if the act of transgression in cyberspace does not cross the thresholds which deem a military response as justified. Moreover, disproportionate retaliation on the part of the victim state, on the other hand, can provoke or instigate an escalation of the conflict. There is a multitude of options before a nation state, varying from economic sanctions to the freezing of financial resources, and political or diplomatic strangulation to the extent of a military response as well.

Retaliatory actions or operations in cyberspace need not be equivalent to the attacks, or confined to similar targets or methods. The underlying principle for deterrence to be effective is that it should threaten unacceptable damage or harm, and the attacker's calculation or understanding of the 'unacceptable' need not be necessarily equivalent to, at par with, or less than the harm or damage inflicted.[58] Also, deterrence in cyberspace cannot be specific or limited to the domain itself.[59] Rather, it deems a more pervasive response, extended to other domains of land, sea, air, or space—that is, more of "cross-domain" deterrence. James A. Lewis has also asserted that deterrence in cyberspace should not be domain limited, and it would require threats in other domains to make it effective. Nevertheless, the cross domain approach increases the risk of miscalculation and an escalation of the conflict.[60]

Experts have also argued in favour of cyber deterrence being cross-domain, and not just restricting responses to the domain itself. This means that cyber deterrence should not only be used as a strategy to dissuade threat actors from attacking in the cyber domain or targeting the information infrastructure. It could also be exercised to threaten an adversary by using cyber means to prevent an undesired action. Amir Lupovici invokes nuclear deterrence to assert this argument, referring to the means a defender adopts to dissuade the adversary from executing an attack, regardless of the nature of the attack, either chemical, biological, conventional, or nuclear thereof.

Nuclear deterrence, therefore, encompasses the threat of the use of nuclear weapons to dissuade a host of attacks, employing the nuclear option as the deterrent threat.[61] On similar lines, cyber deterrence should encompass the threat of the use of cyber means to dissuade the spectrum of threat actors from undesired action(s). Cyber deterrence, in essence, deters attacks in the cyber domain and, at the same time, enables the state to exercise cyber means to deter attacks in other domains as well. Nevertheless, executing and practicing "cross-domain" deterrence is practically quite challenging in cyberspace, since quantifying the damages, losses, and theft in cyber attacks do not fit in the strategic calculus as other traditional threats do. Perhaps, such tenets in cyber deterrence strategy could make deterrence more effective; but it also increases the risks of escalation manifold. Retaliation has to underpin deterrence strategy, the absence of which does not incentivise the adversary to refrain from attack. But, in the case of nation state(s), the scope and methods of retaliation need to be adopted meticulously, with due consideration to political, economic and diplomatic consequences. Both during peace time and under conflict situations, communication is of utmost significance for effective and credible deterrence.

Communication and Signalling

Communication is vital to the process and practice of attribution[62] and is an integral part of the deterrence mechanism. Communicating intentions, thresholds, warnings, threats of punitive measures, or reassurances are necessary for a deterrence strategy to work effectively. In his book *Arms and Influence*, Thomas Schelling underscores the imperatives of effective communication for successful deterrence, using methods either of punishment or denial. It takes a blend of channels, be they political, diplomatic, or military to communicate the intentions, discords, warnings, or even signal to the adversary.

Signalling intentions to the receiver is of utmost importance to the practice of deterrence because, in the absence of such ability, deterrence by punishment runs the risk of being rendered ineffective as well as of being misinterpreted, leading to the escalation of conflict instead of it subsiding.[63] Nation states have to protect their interests, and the actions they would undertake to do so need to be declared. The same principle applies in cyberspace where the interests of nation states are growing day by day, and the need to declare punitive measures, particularly in the case of deterrence by punishment, need to be stated or declared. Communication makes this task seamless. It facilitates the flow of information to both the adversaries and the international community. Communication and signalling forewarns the adversary of the discontent of the defender, and its right interpretation enables the adversary to weight the potential benefits and the costs properly. The adversaries need to know the thresholds and the likely course of measures if the thresholds are breached, while the international community needs to be convinced of the punitive measures adopted.

The indictment of Chinese military personnel by the US judiciary also built diplomatic pressure on the Chinese government to bring them to negotiations, and strike an agreement on the rising intensity of economic and industrial espionage against the USA. The reports from Mandiant, coming out of an independent private entity based on technical evidence, were also part of the communication to persuade the domestic as well as international audience for the retaliatory action, which was broadly restricted to diplomatic manoeuvring. Even statements from intelligence agencies, and the President's office in the case of Sony Pictures hacking (2014) were part of the political signalling being made to the state of North Korea. The Presidential decision in 2017 to elevate the US Cyber Command to a unified combatant command also finds deep political and military signalling to potential adversaries, and underpins the cyber deterrence strategy of the USA.

Intelligence enables nation states to determine the advancing capabilities of their adversaries. This is an essential requirement in order to devise an appropriate response. In the case of nuclear deterrence, as per Glenn Snyder, intelligence, inspection, and satellite surveillance enhanced the ability to detect and determine what the other is doing.[64] In the same context, as per Patrick Morgan, effective communication of the threat is augmented by surveillance.[65] Nation states are known to be conducting espionage operations in cyberspace, which lets them determine the capabilities they would need in the future as part of their deterrence mechanisms. This could further be leveraged to communicate the threat of deterrence by deliberately letting the adversary know about the amassed capabilities. Communication and signalling might be essential for effective deterrence, but acting on the signalling and breach of threshold is also pertinent to the credibility of the threat, for it to be effective in dissuading conflicts of the future. Acting on any breach of redlines has to be limited in scope, severity, and intensity in accordance with existing international legal frameworks.

Necessity/Proportionality

International law enshrines the nation state with the right to respond militarily to an act of war. However, it explicitly restricts the measures to repel or defeat the attack to be based on the principles of necessity and proportionality. The defender or victim state may, however, want to respond disproportionally, but international law categorically limits the responses. These principles seek to reduce the possibility and risk of escalation. In compliance with the Geneva Conventions on the Law of Armed Conflict (1949) and the associated international law, even the Tallinn Manual on the International Law Applicable to Cyber Warfare argues in favour of a proportional retaliatory action in the case of a cyber conflict. It is the responsibility of the nation state to respond judiciously in accordance with these principles to establish the credibility of its retaliatory action among the global community—that is, not to escalate the conflict but rather to ensure the legitimacy of the act. The impacts of cyber attacks could be far reaching, and go well beyond the apparent scope at the beginning of the incident. Damage assessment is hard to make at the first instance. As a result, the retaliatory action might not be commensurate with the damage inflicted. This probably leads to either the erosion of the deterrent effect or the escalation of the conflict. Additionally, the victim needs to make a thorough assessment of the consequences of the retaliatory action, be it political fallout, a diplomatic standoff, a legal confrontation, an economic crisis, or a military contest.

Executing Deterrence by Punishment in Cyberspace: Key Considerations

Article 51 of the Charter of the United Nations preserves the inherent right of nation states to defend themselves against the use of force that threatens their territorial integrity or political independence. In the absence of any known instance of a cyber attack impacting either the territorial integrity or the political independence of nation states, they may not be equivalent to an act of war in the traditional sense. Cyber attacks may not have imposed physical damage as of now; but the possibility and likelihood cannot be outrightly discarded. The deeper integration of information and communication technologies with the infrastructure—such as governance, banking, healthcare, transportation—also leads to the possibilities of real-world catastrophic events owing to cyber attacks. The massive DDoS attacks on Estonia in 2007 crippled its core infrastructure, from banking and transportation to governance. This attack—and also the attacks on Georgia in 2008—did not cause any physical destruction;[66] but they were able to achieve those effects a physical attack could have, invoking the right of Estonia and Georgia to self-defence.

Under the present circumstances, to solicit the right to self-defence and take an appropriate legitimate action in accordance with international law or the UN Charter, is a huge task as there are no agreed upon definitions, defined scope, or measures of assessment in the cyber realm. As a consequence, deterring threats in the cyber domain are quite challenging as compared to the physical domains. In practice, however, the alternative technical and diplomatic approaches—such as robust cyber defences, resilient information systems, as well as confidence building measures and norms are apparently attainable and effective in maintaining stability in cyberspace. There are certain significant issues and risks that arise when deterrence by punishment is exercised as part of a deterrence strategy, which have potential fallout on international stability.

- The primary questions before decision makers when establishing an effective cyber deterrent posture are: establishing thresholds or 'red lines'; communicating these thresholds and threat messages to a diverse set of potential adversaries; and timely detection and swift attribution of the intrusions and attacks. Equally important is the communication of the threat of repulsive action or retaliation, including the capabilities and the intent to put these amassed capabilities into use as and when required. The confidence of intelligence agencies in the attribution of an attack directly impacts the decision of the political leadership. Credible and certain attribution made with higher confidence can push the political leadership to take stringent measures, and even opt for a military backed operation or move diplomatic resources. As of now, most of these questions remain unaddressed and unattended. This partly explains the proliferating instances of attacks aimed at critical infrastructure—such as power and healthcare, governmental databases (Office of Personnel Management of the USA, India's Aadhaar), and even at the election processes of democratic nations, with the USA and French Presidential elections being the prime examples.

- Establishing unambiguous 'red lines' that could define the boundaries of a deterrence policy is essential while articulating a clear distinction between acceptable and unacceptable behaviour pertaining to the national security concerns in cyberspace. Along with this, declaratory measures need to intimate the potential adversary of the likely course of action/retaliation or the quantum of punishment if and when the 'red lines' are crossed. This must influence the cost-benefit analysis of the adversary, with a view to dissuade/discourage the adversary from any such action. To maintain successful deterrence, it is vital to counter the perceived benefits of the adversary with either credible costs for the undesired action, or deny the adversary any benefits from an undesired action.[67]
- As experts argue for "cross domain" deterrence, options in the cyber domain would further integrate deeply with the punitive options available in the physical domain, be it a conventional military response, a nuclear option, diplomatic strangulation, or economic or technology sanctions.
- Akin to the physical domains, compliance with international law is vital for any course of action in cyberspace. Adherence to international law is the foremost criteria, and upholding the principles of necessity and proportionality reinforce the credibility and legitimacy of the deterrence strategy and its tenets for punishment if the nation state is forced to exercise deterrence by punishment. Unfortunately, the

absence of any international agreement or norm regarding the definitions or constitution of responses to cyber attacks, varying perceptions, intangible consequences, and several other factors impair the quantum, severity, and extent of punishment. An unsubstantial response can deteriorate the deterrent effect on the one hand; excessive punishment runs the risk of escalation on the other.

- Non-state actors are playing a vital role, both in the development as well as for the detriment of cyberspace. Non-state actors are exploiting the domain to advance their own goals and motives, which are as diverse as monetary benefits to political objectives. There has been a significant increase in their number and influence on global politics. Amongst them, deterring violent non-state actors in cyberspace is especially challenging because their risk exposure to retaliatory actions is somewhat limited as compared to state actors: they do not possess any physical infrastructure, assets, population, or territory. Moreover, adding to the dismay, these actors exist outside the Westphalian system; so direct communication or structured engagement or negotiations are not possible with them.
- An assessment and recognition of geopolitical symmetry between nation states is also pertinent to devise a response to a cyber attack. Extreme disparity can restrict the options of the weaker state significantly, especially if it cannot afford a conflict spilling over into any of the physical domains. As mentioned earlier in the chapter, Estonia's options were severely constrained in the 2007 incident of DDoS attacks, because Russia, the alleged attacker, had significant geopolitical advantage. Geopolitical clout or dominance can, therefore, be a key determinant in the options available to a nation state when it opts to exercise deterrence by punishment.

Three essentials need to be in place before deterrence by punishment can be exercised. Attribution, being the prime one, is essential for any political leadership to know the perpetrators or the source of the attack. Any response thereafter is made in accordance with the calculations for political benefit, economic leverage, or purely a military operation in self-defence. Beyond the identity, attribution also extends to figuring out the motivations and intentions of the attackers, and whether they have been acting alone or on behalf of any other state or entity. Prompt and certain attribution is indispensable to the potency of a deterrence strategy. The second factor is the applicability of cyber deterrence on the wide spectrum of threat actors. Tailored responses are desirable to deter threat actors with different capabilities and varying risk appetites, be they foreign intelligence agencies or military establishments of a nation state, or just a few criminal syndicates. The last factor is the assessment of damage in terms of the impact of the attack on the critical infrastructure, the society, the economy, the national interest, or any other asset that the nation values the most.

According to the deterrence theory—particularly with respect to nuclear weapons—the undesirable mutually assured destruction resulting from their use serves as a deterrent to prevent nation states from actually putting them to use. However, in the cyber realm, the results of an attack while being devastating may or may not be catastrophic. Furthermore, the lucrative nature of the rewards and the relatively low risk of swift and massive retaliation by the victim make cyber strikes the more viable means of attack. Therefore, the threat of punishment although being the bedrock of cyber deterrence, is not enough to serve as a deterrent alone. The threat of punishment can also be combined with normative frameworks, and the application of the Laws of Armed Conflict (LOAC) can serve as a guideline to address cyber threats and cyber attacks.

Several factors impede deterrence in cyberspace. The asymmetric capabilities and their advantages to target a resourceful entity or a nation state thereof; time lags in analysis or investigations (which easily runs into weeks or months) lead to delayed attribution; the absence of any thresholds of damage which demark the differentiation between an act of war; and the plethora of actors with varying motives, etc. are a few of these factors. In the wake of growing threats to national security, internal stability, economic growth, and even to sovereignty in some cases, nation states are exercising their rights to deter potential adversaries from taking any untoward action in the cyber realm. That also partially explains the reason why cyber deterrence is now the bedrock of cybersecurity strategies of a growing number of nation states.

Cyber Deterrence: The Global Landscape

Deterrence in cyberspace has different notions for different nation states, depending on the varying pedestals of capability, capacity, and intent they stand on to exercise their powers in the cyber domain. However, the concept and its implementation in the form of strategy have gained significant momentum over the last few years. Cyber strategy documents are increasingly making mention about Cyber Deterrence, and laying out the ways and means to achieve it. The broader global developments in this segment are discussed below.

Australia

Australia's Cyber Security Strategy, published in 2016, lays very strong emphasis on growing both defensive and offensive capabilities to such an order that it can deter and respond to any threat of cyber attack, in consistence with the international rules and law. It states that "[i]t is equally important to deter malicious cyber activities by better understanding the threat and bringing the perpetrators to justice."[68] The strategy document does not elaborate on the offensive aspects of deterrence; but it deems strong cybersecurity measures to be pertinent for organisations in order to detect malicious cyber activity and be an effective deterrent by increasing the effort necessary for an attacker to succeed.

The USA

In August 2017, the US President directed the US Department of Defense (US DoD) to elevate the United States Cyber Command (USCYBERCOM), raised in 2009, to a unified combatant command.[69] This move was also seen in the backdrop of the growing number of incidents of hacking targeted at the governmental infrastructure of the USA—attacks on the databases of the Office of Personnel Management, on insurance operators from the healthcare sector, and on the much noted US presidential elections being few of the prominent ones. As a combatant command, the Cyber command would be better equipped and authorised to conduct a host of both defensive and offensive operations to secure the interest of the USA in cyberspace. It is also slated to act as a credible deterrent to dissuade a wide spectrum of hostile actors from targeting the US DoD and other infrastructure that the government deems to be critical.

Both in civilian and military spheres, the USA dominates the arena of cyber deterrence. The USA, armed with technology and a clearly stated intend to practice effective deterrence in cyberspace, targets the threats that could cause wide-scale disruption, destruction, loss of life, and significant economic consequences to its interests. US policy on cyber deterrence is also one of the building blocks of its cybersecurity strategy. In consistence with domestic and international laws, the USCYBERCOM has operational readiness to conduct cyber operations in coordination with other government agencies as appropriate, to deter or defeat strategic threats in other domains. For the US DoD, deterring cyber attacks is an amalgamation of declaratory policy, substantial indications and warning capabilities, defensive posture, effective response procedures, and the overall resiliency of its networks and systems. In addition, it seeks to inculcate strong intelligence, forensics, and indications and warning capabilities to reduce anonymity in cyberspace, and increase confidence in attribution.[70] In 2015, President Obama signed an executive order giving the Secretary of the Treasury authority to issue sanctions against actors that launch cyber attacks on the USA,[71] employing economic means to fortify cyber deterrence strategy.

The UK

The UK plans to strengthen its cybersecurity to such an order that it will be a hard target for all forms of aggression in cyberspace. The National Cyber Security Strategy document of 2016–2021 clearly outlines that the UK has the means to detect, understand, investigate, and disrupt hostile action taken against it, as well as to pursue and prosecute offenders. It also has the desirable means to take offensive actions in cyberspace, if given the option to do so. The strategy explicitly states that defence and protection start with deterrence. It lays deep emphasis on "Offensive cyber capabilities", which primarily involve deliberate intrusions into the opponents' systems or networks with the intention of causing damage, disruption, or destruction. Offensive cyber capabilities will form a part of the full spectrum of capabilities of the UK in the future to deter adversaries, and to deny them opportunities to attack in both cyberspace and the physical sphere. The National Offensive Cyber Programme (NOCP), one of a kind in the world, will equip the UK with the dedicated capability to act in cyberspace. The NOCP will ensure that the UK has appropriate offensive cyber capabilities at its disposal that can be deployed at a time and place of its choosing, for both deterrence and operational purposes, but in accordance with national and international laws.[72]

China

China, with its modernizing military and technology might, is gradually emerging as a key player in the global cybersecurity landscape. China claims to stand for the "peaceful use of the global information space", with the precondition that national sovereignty, interests, and the security of its information domain are protected.[73] President Xi Jinping has asserted clearly about the strengthening of cyber defences and deterrence capabilities to protect key information infrastructure.[74] However, China's National Cyberspace Security Strategy deems the strengthening of the cyber deterrence strategy by other nations as an aggravation of the arms race in cyberspace.[75]

China's position on cyber deterrence can be deciphered from the modernization of its armed forces and the domestic legal framework it has developed recently. The first white paper on China's military strategy in 2015 stressed on the need to shift to "active defence", and emphasized China's commitment to "winning informationized local wars".[76] The white paper also acknowledged China's commitment to build a "cyberforce" with the capability to engage in offensive cyber operations. In accordance with its modernization plan—and possibly in response to global practices—China elevated its major cyber warfare and intelligence-gathering apparatus into the newly established Strategic Support Force, at par with the four services: the army, navy, air force, and the rocket force of the People's Liberation Army (PLA). The SSF includes the 3rd Department of the PLA, which has highly-trained personnel who specialize in network attacks, information technology, code-breaking, and foreign languages; the 4th Department which has the responsibility for military electronic intelligence and electronic warfare; and the 2nd Department which is the traditional military spy service devoted to human spying. Going forward, the SSF is likely to play a vital role in enabling a cyber deterrence strategy for China, comprising forces in the space, cyber, and electromagnetic domains.

Russia

Russia has adopted "strategic deterrence" as a doctrinal approach to leverage a wide spectrum of capabilities and measures which is conceptually different from the Western approach.[77] The Russian understanding of strategic deterrence is much broader, combining both offensive and defensive postures, covering nuclear as well as non-nuclear dimensions, and the use of a host of non-military deterrent tools.[78] The 2014 military doctrine of the Russian Federation emphasized 'non-nuclear deterrence', and underscored a system of military, political, diplomatic, military-technical, and economic measures to prevent an aggression.[79] It lays strong prominence on information space, information technology, and information security.

This, perhaps, makes the Russian deterrence concept more universal—a wholesome approach to deter a wide spectrum of threats utilising all the ways and means.[80] Such an approach was evident in Ukraine, Syria, and the Baltics, where Russia practiced coercion by merging military and non-military forms of influence across nuclear, conventional, and informational (cyber) domains.[81] "Ensuring strategic deterrence" is one of the core objectives outlined in the 2016 doctrine of information security of the Russian Federation.[82]

The Russian armed forces, as per a Ministry of Defence publication, have developed a system to effectively deter, prevent, and resolve armed conflicts in the information space.[83] Reflecting on the role of information and psychological warfare in Russian military thinking, Chekinov and Bogdanov deem it to be essential for gaining superiority, and to depress the armed forces personnel and population of the adversary, both morally and psychologically.[84] In 2011, the Russian Ministry of Defence proposed a document—"Conceptual Views"—on the Activities of the Armed Forces of the Russian Federation in the Information Space,[85] placing information superiority as one of the priority areas for the armed forces from the warfare perspective. While most of Western thought on the subject focuses on the cyber domain *per se*, Russian thinking on the subject is mostly aligned to information space. Subsequent developments have been clearly influenced by this strategic thinking.

Cyber Deterrence: Perspectives from India

India has embarked on the path of intensive digitalisation of the economy and the governance apparatus, notwithstanding the escalating incidence of cyber crime, cyber espionage, and a host of other forms of malwares. India is also risk prone to acts of terrorism, as terror outfits are getting technology savvy and recruiting well-trained professionals.[86] With the changing landscape of Digital India, threat actors and vectors would also grow enormously. Evolving a deterrence strategy and exercising it in true spirit is the need of the hour.

Up till now, India's efforts appear to be directed at strengthening the defences, broadly focussed on deterrence by denial. This encapsulates a legal framework in the form of an Information Technology Act, a Computer Emergency Response Team (CERT-In), a National Critical Information Infrastructure Protection Centre (NCIIPC) as prominent steps to bolster defences against cyber attacks or crimes. India is also pitching its voice in international exercises for norms development, especially at the United Nations Group of Governmental Experts and the London Process. India also hosted the Global Conference on Cyber Space in November 2017, catapulting India into the nucleus of global discourse on cyberspace governance and security. India has a stated Cyber Security Policy, but there is no stated Cyber Security Strategy. It is imperative to analyse the recent developments and the emerging perspectives from the government, academia, and the private sector largely in the context of cyber deterrence. In order to gather the nuances of Indian perspectives on cyber deterrence, a series of interviews were also conducted, drawing in expertise from the government, private sector and think-tanks. The interviews were conducted under the Chatham House Rule.

There is a general agreement among the experts that cyber deterrence is a cornerstone of a competent Cyber Security strategy, and it must, therefore, be practiced in full spirit, scope, and capacity. The first step in this direction has to be a synchronised and concerted national perspective, comprising of ability (in terms of human resources, technology development, infrastructure, etc.) and the political will to exercise cyber deterrence as and when required. Most experts agreed to the fact that, Cyber, as a means of warfare, is extensively being used by nation states, in a stark difference to nuclear weapons which have not been put into use since 1945. The experts converged on the argument that the response to a cyber or physical incident may or may not be restricted to the cyber domain, thus alluding to the "cross domain" tenet of cyber deterrence. Although such a response may not be on the same scale or on the same terms, cyber deterrence should open the path to retaliating against an incident in a different domain altogether.

On the question of the administration of the institution(s) charged with the execution of cyber deterrence, experts strongly agreed that the government should lead these efforts and provide the requisite framework. Since it is infeasible for a single entity to invest in the resources needed and then operationalise a cyber deterrent force, the role of the private sector in terms of investment, human resource, expertise, support, and technology development is vital. The role of the armed forces is also critical to the overall doctrinal and strategic vision guiding a cyber deterrent posture. The armed forces bring in the expertise, operational capability, and a clear mandate to defend the nation from any external threat or transgression. Moreover, they are already playing an important role in thwarting threats in cyberspace. In essence, a cyber deterrence strategy has to interlace the roles, responsibilities, capacities, capabilities, infrastructures, and intellectual capital spread across the government, the armed forces, and the private sector for a coherent and synchronised effort. The principle applies to both the aspects of deterrence by denial and deterrence by punishment, for it to be effective, potent, and credible.

Deterrence by Denial

India's efforts are largely spread in the realm of deterrence by denial, fortifying defences with the help of technology, policy implementation, and public-private partnerships. India's critical sectors of the economy—such as banking, energy, and telecommunication have undergone thorough review and assessments, and these sectors are also implementing the government's commitment towards cybersecurity. The CERT-In and the NCIIPC as the nodal agencies are spearheading preventive efforts, issuing advisories, early warnings and alerts, disseminating vulnerability notes, analysing malwares, cleaning botnets, and reporting and coordinating cyber incidents. The NCIIPC, in particular, aims to reduce the vulnerabilities of critical information infrastructure against cyber terrorism, cyber warfare, and other threats.[87] Both the nodal agencies are investing heavily on capacity building in the form of awareness and training programs. However, amongst the above stated, emphasis is apparently more towards the reactive functions which are rendered inadequate in the face of the rising sophistication and frequency of cyber attacks and malwares.

The lack of indigenous production capability for information technology, communications technology, and the information systems products leads to excessive dependence on foreign products. This elevates the risks of supply chain contamination for both hardware and software, and so also to the security of the information infrastructure *per se*. The deficit in terms of research and development and also in the ability to scale up for the domestic demand, has led the government to aggressively pursue flagship programs such as Make in India. In the short to mid-term, till these programs bear fruit, the government's role as a regulator is to conscientiously ensure the integrity and resilience of the information infrastructure—the key determinants of deterrence by denial.

Based on the interviews conducted with the experts, it is quite evident that overcoming the deficits in capacity and capability is feasible through closer cooperation between the private sector and the government. The government has to take the lead on this in terms of allowing lateral entry across military and civilian organisations. Utilising and tweaking existing models—such as the Territorial Army—for creating an easily "mobilizable" and quick response team to respond to cyber emergencies is also a valuable suggestion. The private sector is also crucial for establishing deterrence through denial, given that most of the information infrastructure—both critical and otherwise—is in the hands of the private sector, and it has the primary responsibility of ensuring resilience to attacks and remediation in the event of a breach, intrusion, or attack. Given that the government has so far not succeeded in enabling effective information sharing through ISACS and the Threat Information Centres, an expert from the private sector also suggested that these functions be outsourced to private companies. Alternatively, individual expertise in the private sector should be identified, and brought into these mechanisms.

Entanglements

As a developing and one of the largest economies in the world, India is experiencing a burgeoning integration of markets, people, research institutions, financial networks, human capital, and a number of other attributes with the other parts of the world. This integration also leads to dependencies in terms of capital, goods, services, and resources. Such dependencies give birth to entanglements, which India already has with the USA, China, Japan, and other major economies and powers across the globe. These entanglements should induce mutual restraint in cyber attacks on each other's infrastructure. However, while entanglement has traditionally been predicated upon mutual interdependence, there is also a fear that such entanglements can turn out to be asymmetric and disadvantageous in the long run in the cyber domain.[88]

Norms

After India's formal acceptance of the multi-stakeholder model of Internet governance, it has proactively begun to participate in global forums for governance, security, and the future of cyberspace. India is an active participant and key player at Internet Corporation for Assigned Names and Numbers (ICANN), and has recently hosted the fifth iteration of the Global Conference on Cyber Space (GCCS) under the aegis of the London Process in November 2017. The GCCS aspires to establish internationally agreed 'rules of the road' for behaviour in cyberspace.

On the pertinent issue of norms development, experts from the private sector gauge them to be an important pillar of deterrence on the international front. If nation states come together on commonly agreed norms, or probably a legally binding arrangement in the future, punishment could possibly be an effective deterrent for malicious cyber behaviour. Cyber conflicts primarily loom at the low end as they are not destabilising as of now. Given the extent of threats before India, ranging from nation states to non-state and state supported actors (Terrorism and Left Wing Extremism), differentiating between these actors is an extremely challenging task. Cyber deterrence strategy has to differentiate between them, and then devise the responses to each of the threat actors. Under such a wide threat landscape, cyber deterrence could further be honed using Artificial Intelligence and Machine Learning, as these respective technologies mature.

Deterrence by Punishment

India's capabilities to conduct offensive operations, to punish the perpetrators in retaliation to a cyber attack are surrounded by ambiguity. There is no clear mandate, doctrine, or policy which authorizes deterrence by punishment. This is unlike the case of the USA and the UK, who have stated tenets of offensive capabilities as part of their respective strategies to deter threats in cyberspace. However, the premier technical intelligence agency, the National Technical Research Organization, is presumed to possess the desired wherewithal to punish an act of aggression in cyberspace, if circumstances and the political requisites deem it fit to act in response.

The sheer absence of a declaratory doctrine, strategy, or policy for exercising deterrence by punishment in the cyber domain constrains the requirements of a credible deterrent in terms of signalling, communicating intentions and thresholds of bad behaviour and fore-warning the potential adversaries of punitive measures. Perhaps well-articulated thresholds (either qualitative or quantitative) for cyber deterrence could be kept in the public domain for consultations and enhancements, on lines similar to India's nuclear doctrine. Swift attribution is critical to the practice of deterrence by punishment; but ambiguities still surround the desirable capabilities to do so. These warrant techno-political solutions.

The interviews with experts rendered differing opinions on developing offensive capabilities, the general opinion being that it led to increasing instability in cyberspace and was also at cross purposes with creating norms and other objectives of the international community—like confidence building.

However, this should not come in the way of academic and research purposes in the interest of national security. It was also pointed out explicitly that India is facing adversaries with major differences in capabilities and vulnerabilities, and deterrent strategies could be effective only if they were tailor-made for the different threat actors. Added to this was the presence of non-state actors and terrorist groups who are increasingly acquiring sophisticated technologies available in underground market places like the Dark web. While deterrence by punishment includes punitive measures—such as diplomatic expulsions, criminal prosecutions and economic sanctions—these do not have much traction in the cyber domain; nor does India have as much leeway as, say the USA, in employing them.

Interestingly, it is not always the capacity of the state to inflict harm on the adversary that enables it to exercise a deterrent posture; it also is the belief of the adversary that the state has such a capacity. Even in a traditional sense, deterrence by punishment features prominently in national security strategies. To establish such a posture, the defending state must first identify the 'red lines' or the 'thresholds' that distinguishes between behaviour it will accept, and behaviour it will strictly punish. Thereupon, it must choose an appropriate course of action or punishment for an actor that still chooses to violate the 'red lines'. Most importantly, it must establish the credibility of its threat of punishment by communicating clearly that it has the ability, resources, and intent to follow through. Deterrence works on the mind of the decision makers; it targets the assets or processes or possessions which the adversary values the most.

NOTES

1. Edward Geist, "Deterrence Stability in the Cyber Age." *Strategic Studies Quarterly*, Vol. 9, No. 4, p. 44.
2. Emilio Iasiello, "Is Cyber Deterrence an Illusory Course of Action?" *Journal of Strategic Security*, Vol. 7, No. 1, 2014, pp. 54–67.
3. Joseph Nye, "Deterrence and Dissuasion in Cyberspace", *International Security*, Vol. 41, No. 3, 2016, p. 46.
4. Robert Jervis, *The Meaning of the Nuclear Revolution*, Cornell University Press: New York, 1989, p. 9.
5. Ibid, p. 10.
6. Erik Gartzke and Jon Lindsay, "Cross-Domain Deterrence: Strategy in an Era of Complexity", Prepared for the International Studies Association Annual Meeting, Toronto, 25–29 March 2014.
7. Patrick M. Morgan, *Deterrence Now*, Cambridge University Press, 2003, p. 3.

8. Liam Nevill and Zoe Hawkins, "Deterrence in Cyberspace: Different Domain, Different Rules", *Australian Strategic Policy Institute*, July 2016, at https://s3-ap-southeast-2.amazonaws.com/ad-aspi/import/SR92_deterrence_cyberspace.pdf? kpqaUl5AVt Xflc31 N8V57RZ2ZnsQ0J1G.
9. Glenn Snyder, *Deterrence and Defense: Toward a Theory of National Security*, Princeton University Press, 2015), pp. 3–4.
10. Ibid.
11. n. 7, p. 1.
12. n. 7, p. 44.
13. Lawrence Freedman, *Deterrence*, Cambridge University Press, 2004, p. 15.
14. Patrick M. Morgan, "Saving Face for the Sake of Deterrence", in Robert Jervis, Richard Ned Lebow and Janice Gross Stein, *Psychology and Deterrence*, Johns Hopkins University Press, 1989, p. 125.
15. Dmitry Adamsky, "From Moscow with coercion: Russian deterrence theory and strategic culture", *Journal of Strategic Studies*, Vol. 41, No. 1-2, pp. 33-60, DOI: 10.1080/01402390.2017.1347872, p. 36.
16. n. 13, p. 60.
17. Patrick M. Morgan, "The State of Deterrence in International Politics Today", *Contemporary Security and Policy*, Vol. 33, No. 1, 2012, pp. 85–107, p. 86.
18. n. 15.
19. Pin Pin Chinese English Dictionary, at http://dictionary.pinpinchinese.com/definitions/t/%E5%A8%81%E6%87%BE-weishe, accessed on 12 August, 2018.
20. Dean Cheng, "Evolving Chinese Thinking about Deterrence: The Nuclear Dimension", *The Heritage Foundation*, 16 August 2017, at https://www.heritage.org/asia/report/evolving-chinese-thinking-about-deterrence-the-nuclear-dimension, accessed on 12 August 2018.
21. James Der Derian, "Cyber-Deterrence," *Wired*, September 1994, http:// www.wired.com/wired/archive/2.09/cyber.deter_pr.html, accessed on 20 January 2018.
22. n. 3, pp. 44-71.
23. Will Goodman, "Cyber Deterrence Tougher in Theory than in Practice?" *Strategic Studies Quarterly*, Fall 2010, pp. 102–135, pp. 105–108.
24. Timothy L. Thomas, "Russia Military Strategy", *Foreign Military Studies Office of the United States Army*, 2015, at https://info.publicintelligence.net/FMSO-RussianMilitaryStrategy.pdf, p. 97.
25. S. G. Chekinov and S. A. Bogdanov, "Strategic Deterrence and Russia's National Security Today", *Military Thought*, January 2012, pp. 21–32.
26. Dorothy Denning, "Rethinking the Cyber Domain and Deterrence", *Joint Forces Quarterly*, No. 77, 2nd Quarter, 2015, pp. 9, 8–15.
27. *Kamal T. Jabbour and E. Paul Ratazzi*, "Deterrence in Cyberspace", in Adam Lowther (ed.), *Thinking about Deterrence Enduring Questions in a Time of Rising Powers, Rogue Regimes, and Terrorism* (Air University Press: Alabama, 2013), pp. 46-47.
28. Dorothy Denning, "Cybersecurity's Next Phase: Cyber Deterrence", *The Conversation*, 13 December 2016, at https://www.scientificamerican.com/article/cybersecuritys-next-phase-cyber-deterrence/, accessed on 22 January 2018.
29. Tim Stevens, "A Cyberwar of Ideas? Deterrence and Norms in Cyberspace", *Contemporary Security Policy*, Vol. 33, No. 1, 2012, p. 148.
30. James Andrew Lewis, "Cross-Domain Deterrence and Credible Threats", *Center for Strategic and International Studies*, May, 2010, p. 1, available at https://csis-prod.s3.amazonaws.com/s3fs-public/170713_Deterrence_Stability_0.pdf?lz1HlfsfMcQGYSynMnhACGXI2 PDjEm0p, accessed 23 January 2018.

31. Joseph S. Nye Jr., "Nuclear Lessons for Cyber Security?" *Strategic Studies Quarterly*, Winter 2011, p. 22.
32. n. 30, p. 4.
33. n. 30, p. 7.
34. Munish Sharma, Voter's Dilemma: Data Leaks and Electoral Interventions, *Institute for Defence Studies and Analyses*, Issue Brief, 22 May 2017, at https://idsa.in/issuebrief/voters-dilemma-data-leaks-and-electoral-interventions_msharma_220517, accessed 23 January 2018.
35. Eric Talbot Jensen, "Cyber Deterrence", *Emory International Law Review*, Vol. 26, p. 782.
36. Frank J. Cilluffo, Sharon L. Cardash, and George C. Salmoiraghi, "A Blueprint for Cyber Deterrence: Building Stability through Strength", *Military and Strategic Affairs*, Vol. 4, No. 3, December 2012, pp. 4–5.
37. Ibid.
38. n. 23.
39. North Atlantic Treaty Organization, "NATO Cooperative Cyber Defence Centre of Excellence", at https://ccdcoe.org/nato.html, accessed on 21 November 2017.
40. "International Strategy for Cyberspace: Prosperity, Security, and Openness in a Networked World", The White House, 2011, available at http://www.whitehouse.gov/sites/ default/files/rss_viewer/international_strategy_for_cyberspace.pdf.
41. Franklin D. Kramer and Melanie J. Teplinsky, "Cybersecurity and Tailored Deterrence", *Atlantic Council Brent Scowcroft Center on International Security*, Issue Brief, December 2013, at http://www.atlanticcouncil.org/images/publications/Cybersecurity_and _Tailored _ Deterrence.pdf, p. 2.
42. Matthew Kroenig and Barry Pavel, "How to Deter Terrorism," *Washington Quarterly*, Vol. 35, no. 2, 2012, pp. 21-36, pp. 22-23.
43. "Cybersecurity's next phase: Cyber-deterrence", *The Conversation*, 13 December 2016, at http://theconversation.com/cybersecuritys-next-phase-cyber-deterrence-67090, accessed on 21 November 2017.
44. Catherine Lotrionte, "Examining the United States' New Norms-Based Approach to Cyber Deterrence: A Better Defense", *Georgetown Journal of International Affairs*, p. 72.
45. n. 28, p. 12.
46. n. 31, p. 33.
47. Martin C. Libicki, "Cyberdeterrence and Cyberwar", *RAND Corporation*, 2009, at https: //www.rand.org/content/dam/rand/pubs/monographs/2009/RAND_MG877.pdf, p. xvi, accessed on 25 November 2017.
48. Tobias Feakin, "Developing a Proportionate Response to a Cyber Incident", *Council on Foreign Relations,* Cyber Brief, 24 August 2015, at http://www.cfr.org/cybersecurity/ developing-proportionate-response-cyber-incident/p36927, accessed on 25 November 2017.
49. David A. Wheeler and Gregory N. Larsen, "Techniques for Cyber Attack Attribution", Institute for Defence Analyses, Paper P-3792, October 2003, p. 1.
50. Thomas Rid and Ben Buchanan, "Attributing Cyber Attacks", Vol. 38, No. 2, 2015, pp. 4–37, p. 4.
51. Robert K. Knake, "Untangling Attribution: Moving to Accountability in Cyberspace", *The Council on Foreign Relations*, Statement Before the Subcommittee on Technology and Innovation, Committee on Science and Technology United States House of Representatives 2nd Session, 111th Congress, 15 July 2010.
52. Clement Guitton, *Inside the Enemy's Computer*, Hurst and Company: London, 2017, p. 25.
53. Ibid, p. 67.
54. n. 50, p. 7.

55. n. 50, p. 10.
56. n. 30, p. 2.
57. James A. Lewis, "Deterrence in the Cyber Age", Center for Strategic and International Studies, 13 November 2014, at https://csis-prod.s3.amazonaws.com/s3fs-public/legacy_files/files/publication/141117_Lewis.pdf, p. 18.
58. Patrick M. Morgan, "The State of Deterrence in International Politics Today", *Contemporary Security Policy*, Vol. 33, no. 1, 2012, pp. 85–107, DOI: 10.1080/13523260.2012.659589, p. 102.
59. n. 30, p. 3.
60. James A. Lewis, "Reconsidering Deterrence in Cyberspace", *Center for Strategic and International Studies*, October 2013, p. 5.
61. Amir Lupovici, "The 'Attribution Problem' and the Social Construction of 'Violence': Taking Cyber Deterrence Literature a Step Forward", *International Studies Perspectives*, Vol. 17, no. 3, 2016, pp. 322–342, p. 325.
62. n. 50, p. 10.
63. n. 2, pp. 54–67.
64. n. 9, p. 59.
65. n. 7, p. 44.
66. Thomas J. Mowbray, "Solution Architecture for Cyber Deterrence", SANS Institute InfoSec Reading Room, 12 April 2010, at https://www.sans.org/reading-room/whitepapers/warfare/solution-architecture-cyber-deterrence-33348, p. 8, accessed on 30 November 2017.
67. Kevin R. Beeker, "Strategic Deterrence in Cyberspace: Practical Application" Graduate Research Project, Department of the Air Force, Air University, Air Force Institute of Technology, Wright-Patterson Air Force Base, Ohio, http://www.dtic.mil/dtic/tr/fulltext/u2/a502250.pdf, p. 15.
68. "Australia's Cyber Security Strategy: Enabling Innovation, Growth and Prosperity", the Australian Government, 2016, at https://cybersecuritystrategy.pmc.gov.au/assets/img/PMC-Cyber-Strategy.pdf, p. 28.
69. Jim Garamone and Lisa Ferdinando, "DoD Initiates Process to Elevate U.S. Cyber Command to Unified Combatant Command", DoD News, 18 August 2017, at https://www.defense.gov/News/Article/Article/1283326/dod-initiates-process-to-elevate-us-cyber-command-to-unified-combatant-command/, accessed on 05 December 2017.
70. "The Department of Defense Cyber Strategy", the US Department of Defense, April 2015, at https://www.defense.gov/Portals/1/features/2015/0415_cyber-strategy/Final_2015_ DoD _CYBER_STRATEGY_for_web.pdf, p. 11.
71. Laura K. Bate, "In Search of Cyber Deterrence", *War on the Rocks*, 24 September 2015, at ttps://warontherocks.com/2015/09/in-search-of-cyber-deterrence/, accessed on 05 December 2017.
72. "National Cyber Security Strategy 2016-2021", Government of the United Kingdom, at https://www.gov.uk/government/uploads/system/uploads/attachment_data/file/567242/national_cyber_security_strategy_2016.pdf, accessed on 05 December 2017.
73. Tang Lan and Zhang Xin, "Can Cyber Deterrence Work?" in Andrew Nagorski (ed.), *Global Cyber Deterrence: Views from China, the U.S., Russia, India, and Norway*, EastWest Institute, April 2010, at https://www.files.ethz.ch/isn/115239/2010-04_GlobalCyberDeterrence.pdf, p. 2.
74. Catherine Wong, "China will boost cyber deterrence powers, vows President Xi Jinping", *South China Morning Post*, 19 April 2016, at http://www.scmp.com/news/china/policies-politics/article/1937224/china-will-boost-cyber-deterrence-powers-vows-president, accessed on 10 December 2017.

75. "National Cyberspace Security Strategy", *China Copyright and Media*, 27 December 2016, at https://chinacopyrightandmedia.wordpress.com/2016/12/27/national-cyberspace-security-strategy/, accessed on 10 December 2017.
76. "China's Military Strategy", Ministry of National Defense: The People's Republic of China, 26 May 2015, at http://eng.mod.gov.cn/Press/2015-05/26/content_4586805.htm, accessed on 10 December 2017.
77. Anya Loukianova Fink, "The Evolving Russian Concept of Strategic Deterrence: Risks and Responses" *Arms Control Association*, July/August 2017, at https://www.armscontrol.org/act/2017-07/features/evolving-russian-concept-strategic-deterrence-risks-responses, accessed on 05 August 2018.
78. Kristin Ven Bruusgaard, "Russian Strategic Deterrence", *Survival*, No. 58, Vol. 4, pp. 7–26, at https://doi.org/10.1080/00396338.2016.1207945, p. 7.
79. "The Military Doctrine of the Russian Federation", December 25, 2014, at https://rusemb.org.uk/press/2029, accessed on 05 August 2018.
80. n. 78, p. 17.
81. n. 13, p. 33.
82. "Doctrine of Information Security of the Russian Federation", The Ministry of Foreign Affairs of the Russian Federation, December 5, 2016, at http://www.mid.ru/en/foreign_policy/official_documents/-/asset_publisher/CptICkB6BZ29/content/id/2563163, accessed on 05 August 2018.
83. "Russian Federation Armed Forces' Information Space Activities Concept", Ministry of Defence of the Russian Federation, at http://eng.mil.ru/en/science/publications/more.htm?id=10845074@cmsArticle, accessed on 05 August 2018.
84. S. G. Chekinov and S. A. Bogdanov, "The Nature and Content of a New–Generation War," *Military Thought*, 2013, pp. 13–25.
85. "Conceptual Views Regarding the Activities of the Armed Forces of the Russian Federation in the Information Space", 2012, English translation by NATO Cooperative Cyber Defence Centre of Excellence, Tallinn, at http://www.ccdcoe.org/strategies/Russian_Federation_unofficial_translation.pdf.
86. Munish Sharma, "Lashkar-e-Cyber of Hafiz Saeed", Institute for Defence Studies and Analyses, IDSA Comment, 21 March 2016, at https://idsa.in/idsacomments/lashkar-e-cyber-of-hafiz-saeed_msharma_310316, accessed on 10 December 2017.
87. "Functions and Duties", National Critical Information Infrastructure Protection Centre, at http://nciipc.gov.in/, accessed on 10 December 2017.
88. Elsa Kania, "Technological Entanglement," Australian Strategic Policy Institute, 28 June 2018, at www.aspi.org.au/report/technological-entanglement.

CHAPTER 3

The Geopolitics of Norms Building in Cyberspace

Whilst norms creation is now being considered within the overall framework of cyber deterrence, cyberspace could be said to have become an arena of geopolitical cooperation and conflict in 1998 when Russia first introduced a draft resolution in the United Nations calling for a discussion on developments in the field of information and telecommunications in the context of international security in the First Committee of the UN General Assembly. In the twenty years since then, there have been a number of efforts to bring order to cyberspace through existing institutions as well as new forums established over the years. Among the more prominent ones are the United Nations Group of Governmental Experts (UNGGE), and the Internet Governance Forum (IGF). There are other forums also that have been created through the efforts of leading global thinktanks and NGOs as well governments. Examples of these are the Global Conference on Cyber Space (GCCS)—also known as the London Process (hosted by respective governments)—the Global Commission on the Stability of Cyberspace (funded by a mix of governments and private Corporations), and the World Internet Conference (also known as the Wuzhan Summit, funded by the Government of China). Other one-off initiatives have included the Tallinn Manual and the Global Commission on Internet Governance.

There are also institutions that are connected with cybersecurity tangentially as a consequence of their functional areas. These include the International Telecommunications Union (ITU) and the Wassenaar Arrangement. These processes can be categorised as quasi-multilateral, fully multilateral, multi-stakeholder, technical, and functional mechanisms. All these forums have had mixed success in their objective of creating the rules of the road for a secure and stable cyberspace. It is instructive to examine some of them in detail to understand their strengths and their shortcomings.

2017 could be said to mark the end of a 20 year long attempt at evolving a consensus on creating a secure cyberspace through a norm setting mechanism. The inability of the 5th Group of Governmental Experts (GGE) set up by the United Nations to arrive at a consensus outcome document came as a shock, given the fact that the previous iterations had been relatively successful in evolving norms, even if at a glacial phase. Many reasons have been ascribed to the collapse of the GGE process; but the fact of the matter remains that the major powers no longer saw it as a useful tool to shape cyberspace according to their interests which had begun to diverge widely. The breakdown of the UNGGE has put a question mark over the whole norm creation mechanism itself. Nonetheless, the deliberations and outcomes of the various GGEs represent a long process of negotiation which should be carried forward. Various other forums are also studied in order to highlight their respective achievements and constraints in order to understand how the norm negotiation process can be further improved and the role India can play in this process.

The UNGGE

The genesis of the UNGGE can be traced to a resolution introduced by Russia in 1998.[1] Russia had proposed the establishment of the GGE to examine the issue of information security, and the first group of governmental experts was set up in 2004 by the First Committee, one of the UN General Assembly's six committees on Disarmament and International Security.[2] However, there was no consensus on the recommendations because of the divergent positions taken by Russia and China on the one hand, and the USA and its European allies on the other, on even the issues to be discussed by the GGE.[3] The points of contention were: i) with regard to the amount of emphasis to be placed on the impact of ICTs on national security, and the threats posed by State exploitation of ICTs for military and national security purposes; and ii) the question of whether the discussion should address issues of information content or whether it should focus only on information infrastructure. There was particular disagreement regarding the claim that trans-border information content should be controlled as a matter of national security. Other areas of disagreement also arose on proposals for capacity building and technology transfer to developing countries. The two-page report to the Secretary General simply noted that "given the complexity of the issues involved, no consensus was reached on the preparation of a final report."[4]

The 2009 GGE

A second group, established in 2009, submitted its report in 2010 with a number of recommendations. This was in large part attributed to a more cooperative approach on the part of the USA.[5] This report laid down the basic elements of securing cyberspace, which have been carried forward by subsequent GGEs. It identified the various malicious actors, the victims as well as major vulnerabilities such as attribution and dual-use potential. It identified confidence building and reducing the risk of misperception resulting from ICT disruptions as the major goals for international cooperation. It also made five recommendations to achieve these goals. These included:

- Further dialogue among States to discuss norms pertaining to State use of ICTs, to reduce collective risk and protect critical national and international infrastructures;
- Confidence-building, stability, and risk reduction measures to address the implications of State use of ICTs, including exchanges of national views on the use of ICTs in conflict;
- Information exchanges on national legislation, national ICT security strategies and technologies, policies and best practices;
- Identification of measures to support capacity-building in less developed countries; and
- Finding possibilities to elaborate common terms and definitions relevant to United Nations General Assembly resolution 64/25.

The 2010 GGE Report recommended dialogue among States on the norms to address the collective risks as well as for protecting critical national and international infrastructure. It also called for measures to promote confidence, stability, and risk reduction. While the report was only stating the obvious, the GGE nonetheless offered governments an important forum to take cognizance of unfolding threats in cyberspace, and for narrowing differences to the extent possible.[6]

The 2011 GGE

A third group was established in 2011 to carry forward the work and recommendations of the 2010 report of the GGE. This group, like the earlier ones, consisted of five permanent members and 10 other member States. Its mandate was to study existing and potential threats in the sphere of information security and possible cooperative measures to address them, including norms, rules, or principles of responsible behaviour of States and confidence building measures with regard to information space.[7] It submitted its report in June 2013.[8]

Among the recommendations, made by consensus, were:

- International law, in particular the UN Charter, is applicable to the cyber-sphere and is essential for an open, secure, peaceful, and accessible ICT environment.
- State sovereignty applies to States' conduct of ICT-related activities and to their jurisdiction over ICT infrastructure within their territory.
- State efforts to address the security of ICTs must go hand-in-hand with respect for human rights and fundamental freedoms.
- States must not use proxies to commit internationally wrongful acts and must ensure that their territories are not used by non-State actors for unlawful use of ICTs.
- The United Nations should play a leading role in promoting dialogue among Member States to develop common understandings on the security of and in the use of ICTs; it should encourage regional efforts, promote confidence-building and transparency measures, and support capacity-building and the dissemination of best practices. The Group recommended regular institutional dialogue with broad participation under the auspices of the United Nations, as well as regular dialogue through bilateral, regional and multilateral forums, and other international organizations.
- The 2010 report recommended further dialogue among States on norms pertaining to State use of ICTs to reduce collective risk and protect critical national and international infrastructure. Common understandings on norms, rules and principles applicable to the use of ICTs by States and voluntary confidence-building measures can play an important role in advancing peace and security.
- Called for common understanding on the application of relevant

international laws and derived norms, rules, and principles of responsible behaviour of States.

- States should encourage the private sector and civil society to play an appropriate role to improve the security of and in the use of ICTs, including supply chain security for ICT products and services.

It also called for exchanges of information and communication between national Computer Emergency Response Teams (CERTs) bilaterally, within CERT communities, and in other forums, to support dialogue at political and policy levels.

The main outcome of this GGE was the acknowledgement that existing international law is applicable to cyberspace, thus paving the way for a universal legal framework. As the UN Secretary General noted in his Foreword to the report, "The recommendations point the way forward for anchoring information and communications technology security in the existing framework of international law and understandings that govern State relations and provide the foundation for international peace and security."

The 2013 GGE

As per the recommendations of the UN General Assembly, the UN Secretary General constituted a new group of governmental experts, with the membership expanded from 15 to 20 in December 2013 and with a mandate to produce a report by June 2015. The group examined existing and potential threats arising from the use of ICTs by States, and considered actions to address them, including norms, rules, principles, and confidence building measures. In addition, the Group examined how international law applies to the use of ICTs by States. It recommended for consideration by States of voluntary, non-binding norms, rules or principles of responsible behaviour of States aimed at promoting an open, secure, stable, accessible and peaceful ICT environment. The 2015 report concentrated on the norms of State responsibility, as well as carrying forward the mandate of the 2013 GGE, and developing a common understanding of the application of international law and norms, elaborating on confidence building measures and capacity building. In deference to the sensitivities of various countries, and even after noting that existing international laws, and by extension, norms and conventions apply in cyberspace, the report fought shy of mentioning international humanitarian law (even as it mentioned its principles such as humanity, necessity, proportionality, and distinction).

The same also applied to the word "self-defence" which was not mentioned even though it was implied in the sentence 'inherent right of States to take measures consistent with international law and as recognized in the UN Charter.'[9]

The recommendations of the 2013 UN GGE, coming as they did at a time when the vulnerabilities of cyberspace were becoming all too apparent, provided the necessary momentum for all stakeholders and for States in particular, to begin discussions on securing cyberspace. The recognition that international law, in particular the UN Charter, is applicable to the cyber-sphere and is essential for an open, secure, peaceful, and accessible ICT environment paved the way for the examination of the international laws that applied. The report tried to be more holistic by including recommendations that State efforts to address the security of ICTs must go hand-in-hand with respect for human rights and fundamental freedoms; that States must not use proxies to commit "internationally wrongful acts"; and States must also ensure that their territories are not used by non-State actors for unlawful purposes. Since these clauses were open for interpretation in more ways than one, there have been considerable differences in interpretation and emphasis on the part of various States. Competing frameworks and regimes have been put forward by different countries along with tools and processes designed to propagate them to likeminded fellow travellers.

This report built on previous reports; but it also reflected the increasing geopolitical strains that increasingly impacted and complicated discussions on cybersecurity as the USA and Western allies tried to create a legal and regulatory framework that imposed costs on countries like China and Russia for carrying out actions detrimental to the former's interests in cyberspace. At the same time, the group tried to play a balancing role by referencing the Shanghai Cooperation Organisation and the International Code of Conduct for Information Security in its report.

The 2015 UNGGE

The 2015 GGE could not agree on a consensus report, thus resulting not just in the failure of the GGE but also dealing a body blow to the process which had coalesced around the rights and duties of States in cyberspace. While earlier GGEs had seen agreement that both were to be derived from existing international law, the 2015 GGE had the crucial mandate of taking the process forward, and spelling out how the laws applied as well as fleshing out norms that would fill in the gaps. Further progress was stymied for a number of reasons, including mutual suspicions on the parts of opposing blocs about the motivations and interests of the other. Whilst the USA in particular wanted progress on delineating the rights of States in responding to cyber attacks, especially in terms of taking appropriate countermeasures, countries that had historical enmity with the USA characterized this as directed against them. The Cuban delegation in its official Statement described this as leading to the militarization of cyberspace.[10] The objections were to the content of Draft Paragraph 34 of the report which was characterised as an attempt "to convert cyberspace into a theatre of military operations and to legitimize, in that context, unilateral punitive force actions, including the application of sanctions and even military action by States [which are] claiming to be victims of illicit uses of ICTs."[11] The delegation saw in this a false equivalence being drawn between the malicious use of ICTs and armed attack. Similar inferences were drawn about the utilisation of International Humanitarian Law since it would again lead to a *de facto* legitimization of "a scenario of war and military actions in the context of ICT." In her remarks, the US delegate accused certain countries of reneging on their declared agreements in previous GGEs instead of taking those agreements to their logical conclusions. In her words,

> A report that discusses the peaceful settlement of disputes and related concepts but omits a discussion of the lawful options States have to respond to malicious cyber activity they face would not only fail to deter States from potentially destabilizing activity, but also fail to send a stabilizing message to the broader community of States that their responses to such malicious cyber activity are constrained by international law.[12]

Indian delegates to the UNGGE also underscored how the process had fallen victim to the divergent political objectives and ideological differences of the major countries. The GGE was seen as being dominated by the Western liberal framework, with countries pushing this framework by paying only lipservice to it in terms of their actions. Even though the GGE consisted of 25 countries, there was major under-representation of developing countries and, consequently, very little recognition of their issues and priorities. The views of the major countries were so far apart on critical questions of cyber war, cyber weapons, and the militarization of cyberspace, that it was quite unlikely there would ever be consensus on it.

The UNGGE ended up falling between two stools in pursuit of the twin objectives of moving forwards on both hard law and soft law (norms). In the process, the progress made on the latter, in terms of creating norms for protecting the core infrastructure of the Internet, norms on State responsibility and on the proliferation of malicious tools, amongst others, also fell victim to its overall failure.

The ITU

The ITU is a UN agency that has three major and specific roles: i) setting technical standards; ii) allocating radio spectrum; and iii) providing technical assistance for capacity building to developing countries. The members of the ITU are a mix of delegations from UN member States, apart from the more than 700 members from the private sector who have been admitted as members after a screening process.[13] The ITU is unique because "it is not only an organizational platform used by member States but also an autonomous norm entrepreneur".[14] This is largely a result of its history.

The ITU was tasked with organising the World Summit on Information Societies (WSIS) in 2003. Subsequently, as per the mandate it received from the WSIS,[15] it set up the Global Cyber Security Agenda (GCA) in 2007 "as a framework for international cooperation to promote cybersecurity and enhance confidence and security in the information society".

The GCA was built on five pillars: Legal Measures; Technical and Procedural Measures; Organizational Structure; Capacity Building; and International Cooperation. A High Level Expert Group (HLEG), consisting of nearly a hundred individuals from various stakeholder organisations, was constituted under the auspices of the GCA. Its report, submitted in 2008, was replete with dissenting views, and exemplified the difficulties of arriving at a consensus in such a controversial area.[16]

The ITU has been proactive in implementing certain aspects of the GCA, including the creation of a Cybersecurity Readiness Index (GCI) wherein countries were rated on certain parameters, such as the existence of national structures for coordinating cybersecurity as well as institutions such as a National CERT. The ITU works in close collaboration with the International Multilateral Partnership against Cyber Terrorism (IMPACT), a body sponsored by the Government of Malaysia. The ITU has also been active in the South Asian region, carrying out a CERT assessment in five South Asian countries—

Afghanistan, Bangladesh, Bhutan, Maldives, and Nepal (ABBMN)—in 2012 under the auspices of the ABBMN ministerial initiative of 2010.

However, the limitations of the ITU were visible in the collapse of the World Conference on International Telecommunications (WCIT) held in December 2012, where a significant number of countries either refused or put their ratification of the resolutions to bring Internet governance within the ambit of the International Telecommunications Regulations (ITR) which had last been revised in 1984.

Russia and China have been at the forefront of proposing various policy measures that ultimately serve to gain more control over their national cyberspace. These efforts were seen first at the ITU Conference in December 2012 where the ITRs were to be re-negotiated/updated. Here, they were joined by Arab countries which also proposed a number of restrictive practices.

Russia proposed, under Article 8, that,

> Member States shall ensure unrestricted public access to international telecommunication services and the unrestricted use of international telecommunications, except in cases where international telecommunication services are used for the purpose of interfering in the internal affairs or undermining the sovereignty, national security, territorial integrity, and public safety of other States, or to divulge information of a sensitive nature.

These proposals were starkly similar to the Russian backed "draft convention on International Information Security", first proposed at an "International Meeting of High-Ranking Officials Responsible for Security Matters" at Ekaterinburg, Russia in September 2011.

Other norm entrepreneurs, such as the Internet Society (ISOC), pushed back noting that this proposal would require "Member States to take on a very active and inappropriate role in patrolling and enforcing newly defined standards of behaviour on telecommunication and Internet networks and in services." ISOC pushed the norm development process since

> such issues are most effectively dealt with by developing national best practices and codes of conduct with appropriate international cooperation. The Internet is built on multi-stakeholder cooperation that includes an important role for governments, but similarly engages the private sector and civil society, through a bottom-up, inclusive process, consistent with the Geneva Declaration of Principles.[17]

The ITU Plenipotentiary meet in Busan, South Korea in 2014 again saw a determined effort by a set of countries, including China, Russia, and the Arab countries, to have the ITU play a greater role in Internet governance. Among the resolutions moved were the ones calling for all governments to "have an equal role and responsibility for international Internet governance" (Resolution 102),[18] "protect their Internet Protocol-based networks from unlawful surveillance" (Resolution 101),[19] and asking member States to refrain from "using ICTs involving the extraterritorial interception and monitoring communications in a way which violates the privacy of communications and users' personal data protection" (Resolution 130).[20]

India put forward a proposal on ITU's role in Realizing Secure Information Society (Proposal 98).[21] This called for a 'traffic routing plan' to "effectively ensure the traceability of communication" citing cybersecurity concerns arising out of the fact that there was no way to trace the origin of malicious data and traffic. The weaknesses in the current system network architecture made it easy to camouflage the "identity of the originator of the communication" while random IP address distribution made the "tracing of communication difficult". Other issues with the current system included insecurities from traffic originating and terminating in the same country (domestic traffic) invariably routing through international networks and being susceptible to interception. This proposal was met with resistance with countries, noting that it would require substantial changes to the structure of the Internet, and was beyond the ITU's mandate. It was suggested that the proposal was better suited for discussion at a multi-stakeholder forum such as the Internet Governance Forum.

Given the multilateral nature of this forum, it may be seen that blocs of countries play a role both in putting forth proposals as well as in blocking them. Other than the fact that it is still perceived as a body largely dealing with telecommunications and not cybersecurity, ITU events also occur all too rarely to play an effective role in cybersecurity. There are also already in existence a number of bodies dealing with technical issues such as the IETF; the ITU's role would be overlapping with theirs. For the foreseeable future, the ITU would be playing a tangential role, though countries supporting a role for the ITU will, no doubt, still persist in their efforts to make ITU a nodal organisation for cybersecurity.

The Internet Governance Forum (IGF)

The Internet Governance Forum (IGF) was established by the 2005 WSIS, which was held in Tunis. It authorised the UN Secretary-General to create a mechanism to enable multiple stakeholders to discuss Internet governance. While the emphasis of the mandate was on Internet governance, cybersecurity was also covered in the mandate, viz: discuss public policy issues related to key elements of Internet governance in order to foster the sustainability, robustness, security, stability, and the development of the Internet.[22]

Its mandate was renewed for a further five years in 2011 by a resolution of the UN General Assembly.[23] Its existence was again the subject of the ten year review of Internet policies laid out at the World Summit on the Information Society (WSIS) at the WSIS+10 High-Level Meeting convened by the General Assembly in December 2015. The mandate of the IGF was again renewed for another ten years.[24] While the outcome document covered four main areas—Internet accessibility, human rights and free speech, Internet governance, and cybersecurity—it is striking that the section on cybersecurity only referred to threats from cyber criminals and terrorists.

A report brought out by the IGF in 2015 highlighted the problems faced in ensuring that there was equitable participation from across the world in the multi-stakeholder model that has become prevalent not just in Internet governance but also in cybersecurity.

A region wise pie chart of participation between 2006 and 2010 shows a disproportional number of participants from the developed countries.

Figure 3.1: Participation by Geographic Region 2006-10

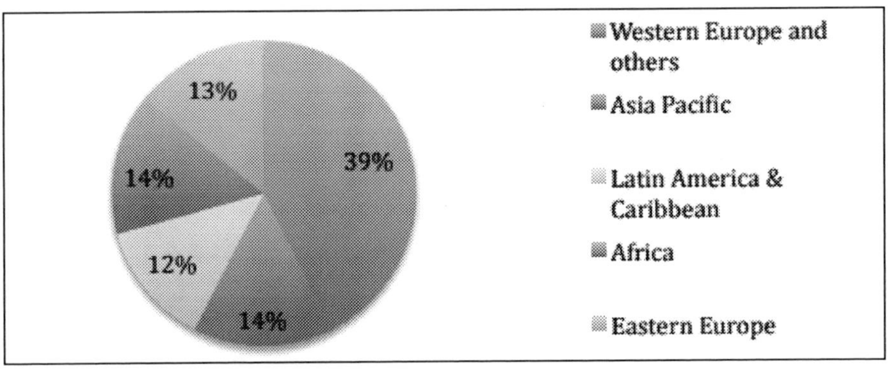

Source: Internet Governance Forum.

This become even more pronounced in the years from 2010–2015[25]

Figure 3.2: Participation by Geographic Region 2010-15

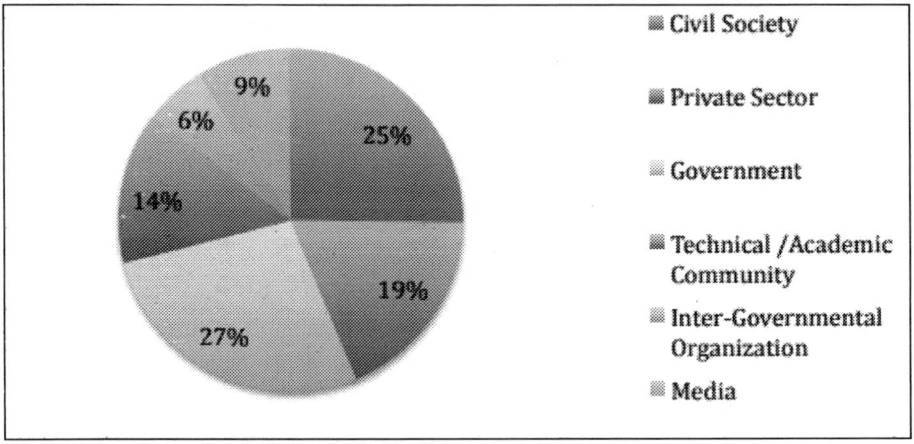

Source: Internet Governance Forum.

Indian attendees have reiterated the fact that, their apparently open nature notwithstanding, these bodies are exceedingly difficult to engage with on a sustained basis since the manpower and resources required are not available. Even entities with these resources, from the government to private companies, are not able to do so at an institutional level; participation also depends on individual interest within these entities. Though there has been a rise in participation by civil society organisations, they have largely been on the basis of external fellowships and funding, which comes with its own sets of issues.

A detailed study undertaken by the Centre for Communication Governance of the National Law University of Delhi on the Indian engagement at the IGF also reiterated the same points. According to the study, it is not just that the participation is low; there are also fewer interactions from those present.[26]

London Process/GCCS

The London process began as a conference on cyberspace hosted by the British Foreign Office following a proposal by the British Foreign Minister, William Hague, at the Munich Security Conference in 2011 for an international meeting to discuss "rules of the road" in cyberspace. This was in response to efforts by

Russia and China to develop an alternative model for cyberspace governance that stressed national sovereignty in cyberspace, and which had culminated in the tabling of an "International Code of Conduct for Information Security" at the United Nations in 2011 by China, the Russian Federation, Tajikistan, and Uzbekistan.

The seven "principles" proposed by William Hague for the conference to provide direction to the conference included the following:

- The need for governments to act proportionately in cyberspace and in accordance with national and international law;
- The need for everyone to have the ability—in terms of skills, technology, confidence and opportunity—to access cyberspace;
- The need for users of cyberspace to show tolerance and respect for diversity of language, culture and ideas;
- Ensuring that cyberspace remains open to innovation and the free flow of ideas, information and expression;
- The need to respect individual rights of privacy and to provide proper protection to intellectual property;
- The need to work collectively to tackle the threat from criminals acting online; and
- The promotion of a competitive environment which ensures a fair return on investment in network, services and content.

In keeping with the above, the conference themes were economic growth and development, social benefits, safe and reliable access, international security, and cyber crime. In his opening Statement, William Hague identified the objectives of the conference thus:

> We want to widen the pool of nations and cyber users that agree with us about the need for norms of behaviour, and who want to seek a future cyberspace based on opportunity, freedom, innovation, human rights and partnership, between government, civil society and the private sector.[27]

With a number of senior level speakers representing government, the private sector, academia, and NGOs, the conference did succeed in its objective of creating an alternate narrative, and Britain ensured that the momentum would be kept up by enlisting Hungary and South Korea to hold the next two conferences.

The Budapest Conference in the following year saw European countries highlighting the human rights aspects of cybersecurity, based on their characterisation of Internet freedom as a fundamental right. That drew an acerbic reaction from China, with the Chinese representative asking whether he was at a human rights conference or a cybersecurity conference. In his speech at the conference, the Chinese representative reiterated the principle of "network sovereignty", and highlighted the need to balance the free flow of information against the potential for threats to national security and social order. He also called for equal rights in managing the Internet, and equitable distribution of the critical resources of the Internet.[28] However, the British continued with their stewardship of what was now becoming known as the London Process by announcing the establishment of a Centre for Cyber-Security Capacity Building at a cost of 2 million pounds.

The Korean iteration was meticulously conducted in 2013, with the hosts preparing a Statement before the conference, and which was subsequently discussed and agreed to by the participants. This made the conference more outcomes oriented, and easier to manage. The "Seoul Framework for and Commitment to an Open and Secure Cyberspace" added a sixth theme of "Capacity Building" to the existing five. This followed on from the 2013 Report of the UN Group of Governmental Experts (UNGGE), which also highlighted capacity building, and in fact, it seemed largely in lockstep with the recommendations in the GGE Report, providing further ballast to the latter.

While cast in the multi-stakeholder mode, there was an increasing tilt towards State participation with both the Budapest and Seoul Conferences being criticised for being too State centric as well as being dominated by Western countries, with little participation from the developing world. Although the Seoul Conference in 2013 had as many as 43 participants at the official Ministerial level, the Snowden revelations took off much of the sheen of the Seoul Conference, and detracted from the main objectives of gaining consensus on contentious issues.

The process itself seemed to have begun to lose steam when it was announced that the next Conference would take place only after a gap of two years, to be hosted by the Netherlands. In the intervening two years, the Dutch government expended a considerable amount of energy and resources on shaping an agenda and gathering support for a successful outcome. The stumbling blocks of low multi-stakeholder participation and low participation from the developing world was sought to be mitigated through support for a series of regional conferences to provide inputs to the larger summit. Thus, no less than 13 preparatory events were held in different parts of the world on varied issues related to cyberspace.

The various sessions of the Conference were centred on the main themes of freedom, growth, and security. The key objectives were: i) support practical cooperation in cyberspace; ii) promote capacity building and knowledge exchange in cyberspace; and, iii) discuss norms for responsible behaviour in cyberspace. As many as 21 countries sent representatives at the Ministerial level, although Russia and China played a very low key role, with hardly any representation other than by in-country diplomats or from neighbouring countries. The majority of the participants, on the whole, were government officials from various countries. The outcome Statement of the Conference was in the form of a Chairman's Statement, which summarised the two days of discussion. The launching of a Global Forum of Cyber Expertise (GFCE) was one of the tangible outcomes of the Conference. Its 42 members included 29 countries, seven private-sector entities, and six intergovernmental organizations.

The next GCCS was to be held in Mexico; but it was shifted to India after Mexico expressed its inability to host the Conference. The Indian iteration sought to highlight "an inclusive cyberspace", with a "focus on policies and frameworks for inclusivity, sustainability, development, security, safety & freedom, technology, and partnerships for upholding digital democracy, maximizing collaboration for strengthening security and safety, and advocating dialogue for digital diplomacy."[29] There were the usual criticisms about lack of participation by stakeholders form civil society and academia which further served to highlight continuing problems with implementing the concept of multi-stakeholderism. Among the highlights of this GCCS was the release of the Delhi Communiqué on a Global Agenda for Cyber Capacity Building by the GFCE which looks to be one of the successful outcomes of the London Process. However, coming on the heels of the collapse of the UNGGE, the GCCS seems to have lost some of its relevance, and this can be seen in the fact that unlike the previous iterations, no country was announced as the host of the next edition.

SCO

The Shanghai Cooperation Organisation was set up in 2001 and has 8 full members, 4 observers, and 6 dialogue partners.[30] Its objectives are

> strengthening mutual confidence and good-neighbourly relations among the member countries, promoting effective cooperation... making joint efforts to maintain and ensure peace, security and stability in the region, and moving towards the establishment of a new, democratic, just and rational political and economic international order.

The SCO has an active cybersecurity initiative, with programs under its Regional Anti Terror Structure (RATS), including cyber anti-terrorism exercises. An Internet Expert Group was setup in 2013 "to fight against online activities of terrorism, separatism and extremism". In addition to cooperation among the member countries, the SCO has also been proactive in placing an International Code of Conduct for Information Security before the General Assembly of the United Nations. While an initial code was tabled in September 2011, an updated draft was tabled in February 2015.

The purpose of this Code was to identify States' rights and responsibilities in information space, promote their constructive and responsible behaviours, and enhance their cooperation in addressing the common threats and challenges in information space, so as to ensure the ICTs (including networks) to be solely used to the benefit of social and economic development and people's well-being, and consistent with the objective of maintaining international stability and security.

Each State that signed the Code essentially voluntarily agreed to the following:

- To comply with the UN Charter and universally recognized norms governing international relations, which enshrine, inter alia, respect for the sovereignty, territorial integrity, and political independence of all States, respect for human rights and fundamental freedoms, as well as respect for the diversity of history, culture, and social systems of all countries.
- Not to use ICTs (including networks) to carry out hostile activities or acts of aggression, and pose threats to international peace and security. Not to proliferate information weapons and related technologies.

- To cooperate in combating criminal and terrorist activities which use ICTs (including networks), and curbing the dissemination of information which incites terrorism, secessionism, and extremism, or undermines other countries' political, economic, and social stability as well as their spiritual and cultural environment.
- To endeavour to ensure the supply chain security of ICT products and services, prevent other States from using their resources, critical infrastructures, core technologies and other advantages, to undermine the right of the countries which accepted this Code of Conduct, to the independent control of ICTs, or to threaten other countries' political, economic and social security.
- To reaffirm all States' rights and responsibilities to protect, in accordance with relevant laws and regulations, their information space and critical information infrastructure from threats, disturbance, attacks, and sabotage.
- To fully respect rights and freedom in information space, including rights and freedom of searching for, acquiring, and disseminating information on the premise of complying with relevant national laws and regulations.
- To promote the establishment of a multilateral, transparent, and democratic international management of the Internet to ensure an equitable distribution of resources, facilitate access for all, and ensure a stable and secure functioning of the Internet.
- To lead all elements of society, including its information and communication private sectors, to understand their roles and responsibilities with regard to information security, in order to facilitate the creation of a culture of information security, and the protection of critical information infrastructures.
- To assist developing countries in their efforts to enhance capacity-building on information security and to close the digital divide.
- To bolster bilateral, regional, and international cooperation, promote the United Nations' important role in the formulation of international norms, the peaceful settlement of international disputes, and the improvement of international cooperation in the field of information security, and enhance coordination among relevant international organizations.
- To settle any dispute resulting from the application of this Code through peaceful means, and refrain from the threat or use of force.

The updated draft had new sections emphasizing the role of States in Internet governance, and the need for confidence building measures, thus bringing it in sync with the UNGGE.[31] An additional section called on the signatories

> not to take advantage of its dominant position in the sphere of information technology, including, inter alia, dominance in basic resources, critical infrastructures, core technologies, products and services of ICTs and information and communications networks, to undermine other countries' right of independent control of ICT products and services, or to threaten other countries' political, economic and social security.[32]

The SCO perspectives on cybersecurity largely reflect the views of its most dominant members, China and Russia. The SCO has been effectively used as a forum to give legitimacy to their perspectives, especially cyber sovereignty, and to push them in international forums such as the United Nations. More recently, it has sought to give a practical push to cooperation on information security, couching it in the language of anti-terrorism activities conducted over the Internet. This is at par with the SCO conceptualization of information security to be led by States, and to include threats from content. The first such anti-terror exercise was carried out in October 2015, while the latest round was carried out in December 2017.[33]

The Wassenaar Arrangement

The Wassenaar Arrangement on Export Controls for Conventional Arms and Dual-Use Goods and Technologies, as the name suggests, is an export control instrument agreed to by 41 countries in 1996, with the objective of "promoting transparency and greater responsibility in transfers of conventional arms and dual-use goods and technologies." This is done through two lists: i) a munitions list, which largely consists of conventional military equipment; and ii) a dual-use Goods and Technologies List of items that can have both a civilian and military use listed under various categories, ranging from avionics to navigation and propulsion.

The move to include cyber products in the Wassenaar Arrangement came from an unlikely coalition of human rights activists and technologists increasingly concerned by the use of intrusion software to target activists in totalitarian countries. Intrusion tools such as Finfisher from Gamma Group

International, or the Remote Control Software (RCS) from the Hacking Team were found to have been used by oppressive regimes to locate anti-government protestors by tracing their digital footprints. According to campaigners, these companies operated completely in the dark, with no oversight despite the fact that these technologies could be reverse-engineered, and proliferated easily once they fell in the hands of terrorists and criminals. Discussions were initiated by the UK in 2012 in the working, and a set of proposals were considered and agreed to in the plenary meeting of December 2013. The areas covered included "surveillance and law enforcement/intelligence gathering tools and Internet Protocol (IP) network surveillance systems or equipment, which, under certain conditions, may be detrimental to international and regional security and stability."[34]

The US Bureau of Industry and Security (BIS) published rules[35] as they apply to cyber technologies in May 2015. While the focus of the updated Wassenaar Arrangement Control was on such surveillance products, the proposed BIS rules were criticised for going beyond IP surveillance, and placing greater emphasis on licensing the export of software classified as intrusion malware and intrusion exploits as well as software that could conceivably go into the production of such malware. The violation of these rules could result in a 20 year jail term and $1 million fine. The software research community in the USA protested against the new regulations once they were opened for comments, particularly since they were not consulted beforehand. This went against the grain of multi-stakeholderism touted by the USA. Their objections were that these rules were so vague as to make even legitimate research into vulnerabilities a possible violation. This would have a chilling effect on such research and would, in fact, have the opposite effect of prolonging vulnerabilities in the software. Digital rights advocates, such as the Electronic Frontier Foundation, highlighted the various anomalies in the BIS rules as well as the subsequent explanatory notes and clarifications issued by the BIS.[36]

Information security companies such as Symantec and Metasploit noted that the rules would severely harm their use for legitimate research purposes. Symantec noted that the rules would effectively have the following repercussions: i) restricting access to legitimate cybersecurity technologies and testing tools across borders—even among security professionals who work for the same company; ii) curtailing research into cybersecurity vulnerabilities, as researchers would be hindered from testing networks and sharing technical information across borders; iii) limiting cybersecurity threat information sharing and collaboration on cybersecurity risks, both within security companies and with customers and industry partners, as information would be deemed "exported" if it is shared with any non-US persons even if they are physically located in the US and employed by a US company.

In point of fact, the BIS rules, if implemented, would deal a body blow to the entire security research ecosystem, affecting everything from bug bounty programs to cross-border research on vulnerabilities. If anything, this would only make it more difficult for genuine security researchers to carry out legitimate research activities into defensive products while creating an underground market for offensive products. This also creates a problem for information security companies, some of which have as much as 70 per cent of their workforce outside the USA.[37]

As far as intrusion exploits are concerned, it has been pointed out that the National Security Agency has also been in the market for zero-day exploits, and would effectively get free access to these exploits via the new rules. The rules would also have the intended or unintended effect of driving the production of cybersecurity software into the hands of those who already have experience with export control rules, namely the large military equipment producers who are already buying smaller companies in their quest to dominate the cybersecurity market.

It is not just US companies and researchers that are bothered about the new rules. As early as June 2014, an inter-ministerial panel of the Indian government was formed to study the impact of the new rules on procurement of software and cybersecurity products.[38] The security of such products is impacted by these rules even post-purchase, since even something as commonplace as auto-updating of browsers is deemed illegal. There might well be a chilling effect on the procurement of cybersecurity products from US companies given that a preliminary reading of these rules indicates that the source code of sensitive products should be examined by the relevant authorities prior to export.

The United States has effectively used a mix of national export controls laws and multilateral export control regimes to control and regulate the flow of technology that could be weaponised. International regimes include the Nuclear Suppliers Group (NSG), The Australia Group (AG) (for regulation of chemical and biological technology), the Missile Technology Control

Regime, and the Wassenaar Arrangement on Export Controls for Conventional Arms and Dual-Use Goods and Technologies. Given that these regimes come with many layers of red tape and bureaucratic oversight, the recent expansion of the Wassenaar Arrangement to cover surveillance and intelligence gathering software was met with concern in the information security community.

Companies would still sell these technologies, demanding higher prices for their services pointing to these restrictions, and giving diluted service level agreements. It could even be said that Western governments are being hypocritical by taking the moral high ground despite their own intelligence agencies making use of these same technologies against their citizens.

The Tallinn Manual

The Tallinn Manual was an effort by a group of international lawyers brought together under the auspices of the NATO Centre of Excellence in Tallinn, Estonia to debate the applicability of existing international laws to cyberspace since the developments had outpaced the evolution of the law.[39] Some of the sources used while creating the Tallinn Manual were the UN Charter, the Military Manuals of NATO countries, the Geneva Convention, various International Treaties, and International Law books and articles, etc. The Manual tried to examine various issues of customary international law in the context of cyberspace. Chief among these were: i) what constitutes "use of force" in cyberspace?, and ii) the violation of Article 2(4) of the UN Charter in the context of cyberspace.

Article 2(4) allowed for the use of force in self-defence; but this became valid only in the event of a "cyber armed attack". By the classic definitions of an armed attack, none of the events so far (including Stuxnet) had crossed the threshold in terms of scale and effects since injury or death, or destruction of property, was the measure of an armed attack.[40] Thus, the Tallinn Manual postulates that there must be an armed attack before the State can respond under the right of self defence.

Other issues covered include neutrality and lawful targeting. The protections afforded to civilians were also covered comprehensively. Civilians enjoyed the protections afforded by the laws of armed conflict so long as they did not directly participate in the conflict. Direct participation included: i) conducting cyber attacks; and ii) any actions which made possible specific attacks (e.g., identifying vulnerabilities or designing malware specifically to take advantage of particular identified vulnerabilities). Indirect participation included: i) designing malware without the specific intention of it being used in the conflict; and ii) maintaining computer equipment generally, even if such equipment is subsequently used in the hostilities.[41]

The Tallinn Manual was largely disavowed by its sponsors. The US position on adapting the laws of war to cyberspace, as a case in point, was that cyberactivities may, in some circumstances, constitute the use of force under Article 2(4), and may lead to an appropriate response. The relevant factors that need to be taken into consideration include context, actors, effects, target, and location. Cited examples included cyber attacks on nuclear installations, attacks against critical infrastructure causing disruption in populated areas, etc.

A second expanded iteration of the Tallinn manual was brought out in 2017, including "cyber operations" or actions below the threshold of war within its ambit. While the former contained 95 rules as developed by the expert group, the latter had 154 rules on cyber operations. It tried to address many fundamental issues, including sovereignty, and actions in cyberspace that could be considered a violation of the sovereignty of a State. It also looked at the responsibilities of States, particularly on the need for due diligence by States to ensure that malicious cyber activities did not take place from their territories.[42]

Even as the authors have been at pains to suggest that the Manual is only to serve as a guide on these issues and to provide a framework within which to approach them, there have been criticisms regarding the framework itself, and the contradictions between the various rules as well as the ambiguities thereof. Sovereignty and Jurisdiction are incompatible, with the former being based on territory, and the latter, as adjudicated by various courts, based variously on the location of the data itself, or based on the citizenship of the owner of the data. The Tallinn Manual takes the position that Jurisdiction is based on the location of the data, but does not satisfactorily address the fact that data itself is movable.

World Internet Conference (China)

The World Internet Conference was begun in 2014 as an initiative of the newly setup Cyberspace Administration of China. The official theme of the Conference was "An Interconnected World Shared and Governed by All". A large number of international experts, government officials, heads of Western companies, and others were invited for the Conference. The aim of the Conference was to present China's view of cyberspace governance as a viable alternative to the prevailing US-centric system. Many global CEOs were present, despite the fact that the services of their companies were blocked in China.[43] The Conference ended on a controversial note since the attendees refused to sign the Wuzhen Declaration.

The Declaration reads as follows:

> The First World Internet Conference was successfully held in Wuzhen, Zhejiang Province from November 19 to 21, 2014. Participants in the Conference acknowledge that the Internet is increasingly becoming a leading force of innovation-driven development and is powering economic and social progress. The Internet has turned the world into a global village and made the international community a highly interdependent community of common destiny. While enjoying rapid development, the Internet has posed new challenges to national sovereignty, security and development interests, which require the international community to meet urgently and seriously, expand consensus and strengthen cooperation.
>
> We call on the international community to work together to build an international Internet governance system of multilateralism, democracy and transparency and a cyberspace of peace, security, openness, and cooperation.
>
> First, enhance cyberspace connectivity. We should accelerate efforts to build Internet infrastructure, increase bandwidth, break information barriers and remove the information gap, so as to ensure that more developing countries will benefit from an interconnected information expressway.
>
> Second, respect Internet sovereignty of all countries. We should respect each country's rights to the development, use and governance of the Internet, refrain from abusing resource and technological strengths to violate other countries' Internet sovereignty, and build an Internet order of equality and mutual benefit.
>
> Third, jointly safeguard cybersecurity. We should actively cope with challenges to cyberspace security and reject all forms of cyber attacks and Internet theft. We should work together to fight cyber crimes,

> protect individual privacy and information security, and safeguard the legitimate rights and interests of citizens.
>
> Fourth, jointly fight cyber terrorism. We should work for the establishment of an international cooperation mechanism against cyber terrorism to fight cyber terrorism together and destroy all dissemination channels of information of violent terrorism.
>
> Fifth, advance development of Internet technology. We should strengthen research and development, dissemination and cooperation on cloud computing, big data and Internet of things, carry out personnel exchanges, and promote more extensive and secure application of sophisticated Internet technologies.
>
> Sixth, vigorously develop the Internet economy. We should improve cyberspace trade rules, step up cross-border e-commerce cooperation, facilitate customs clearance and logistics, expand information consumption, and quicken steps to form a global Internet market.
>
> Seventh, widely spread the positive energy. We should carry forward and promote fine cultures and produce more digital cultural products of high quality, in order to meet people's cultural needs and give a sense of belonging to mankind in cyberspace.
>
> Eighth, dedicate to the healthy growth of young people. We should strengthen the protection of minors online, crack down on the spread of pornography and violence, and make sure that the Internet does not damage the future of mankind.
>
> Ninth, work for a cyberspace shared and governed by all. Following the principles of mutual respect and mutual trust, we should set up a regular cooperation mechanism and communication platform of cyberspace, so as to deepen communication, pursue common governance and realize win-win results, and jointly create a bright future of the Internet.

Probably, consequent to the failure of the first Conference, the focus of the second conference was on State representatives. Premiers invited and attending included the Prime Ministers of Russia, Pakistan, Kazakhstan, and Kyrgyzstan. The theme of the second conference was "Building a Cyberspace Community of Shared Destiny".

In his remotely delivered speech, President Xi Jinping made a number of pointed references to the USA. Among the points that he made were:

(a) Cyber surveillance, cyber attacks and cyber terrorism have become a global scourge;
(b) Cyberspace should not become a battlefield for countries to wrestle with one another, nor should it become a hotbed for crime;
(c) Double standards should not be allowed in upholding cybersecurity;
(d) We cannot just have the security of one or some countries while leaving the rest insecure, and countries should not seek the so-called absolute security of itself at the expense of the security of others; and
(e) All countries should work together to contain the abuse of information technology, oppose cyber surveillance and cyber attacks and reject a cyberspace arms race.

One of the major outcomes of the 2nd Conference was the announcement of a 31 member High-Level advisory Committee to the WIC, including Paul Wilson (General Manager of APNIC), Eugene Kaspersky, Shaukat Aziz (former Prime Minister of Pakistan), and Bruce McConnell (of the East-West Institute).[44] As Adam Segal of the Council of Foreign Relations and an attendee noted, the conference is not so much about content as it is about symbolism. The fact that the Chinese President delivered his speech in person indicates the importance China attaches to driving the discussion on cyberspace, as well as making its perspective known. As he put it, "China is no longer the outside voice at venues organized by others, but has its own platform to promote a competing vision."[45]

Conclusion

In 2011, the academician Tim Maurer authored a paper titled "Cyber-norm emergence at the United Nations" in which he examined the process of norm formation, and the roles of UN-affiliated bodies, such as the International Telecommunications Union and the Internet Governance Forum, in propagating the norms. Maurer largely drew on the work of Michael Barnett and Martha Finnemore to explain the role of "norm entrepreneurs" at the initial stage of the norm life cycle. The role of the norm entrepreneurs is persuasion through organizational platforms. In the years since 2011, the role of such norm entrepreneurs has only grown, as has also the organizational platforms. Their role has been further legitimized through the adoption of the multi-stakeholder model with a role to play for all, as opposed to the multilateral model in which only States were recognized as actors.

While the multi-stakeholder model seemingly democratizes and provides equal opportunity for all in decision-making, in practice its biggest flaw is that it gives a bigger voice to groups of actors who have the wherewithal to be present at the table. This makes many of the organizational platforms for norm creation only accessible to actors from the richer countries. The poorer countries are doubly disadvantaged in not having the resources to fix the vulnerabilities they are exposed to for lack of knowledge and by not being present at the table and having their voices heard.

This also applies to countries that are outside the charmed circle—such as China and Russia—who have then gone on to create their own organizational platforms to propagate alternate norms. China has a multi-pronged strategy in creating a platform like the World Internet Conference to propagate its thinking. While other platforms, like the Shanghai Cooperation Organisation, do exist, they are constrained by the presence of other actors who might not necessarily have the same point of view. Within China, various new organisations, such as the Chinese Culture Institute of Internet Communication, are being created to actively engage in domestic and international forums. Traditionally, norm entrepreneurs have derived their legitimacy from working at the domestic level to make norms a part of State policy that were then internationalized. China has, at the same time, actively engaged with other platforms—like the ASEAN Regional Forum—where it has co-sponsored a number of workshops with Malaysia. Its imprint is clearly seen in the final draft of the ASEAN Work plan which has references to cultural diversity and national sovereignty.

While there seems to be a rough consensus on the multi-stakeholder mechanism, it is ironically those countries that championed the model that now seemed to have lost enthusiasm for it. Going forward, while many States still see utility in evolving norms, the focus seems to have shifted from negotiating norms with adversaries to shaping norms by like-minded countries, which sets the stage for norm competition in cyberspace. These developments could result in the eventual fragmentation of cyberspace, which would be a setback for developing countries that are just beginning to enjoy the benefits of digitalization.

Part of the problem is that the developing countries neither have the heft nor the internal and external capacities to make their voice count in cyberspace. Efforts need to be made to improve the capacities of all stake holders in less-privileged countries to improve their cybersecurity as well as their ability to contribute effectively to the discussion.

NOTES

1. "Developments in the Field of Information and Telecommunications in the Context of International Security", United Nations, at http://www.un.org/disarmament/topics/informationsecurity/.
2. Eneken Tikk-Ringas, "Developments in the Field of Information and Telecommunication in the Context of International Security: Work of the UN First Committee 1998-2012", ICT for Peace, 2012, at http://www.ict4peace.org/wp-content/uploads/2012/08/Eneken-GGE-2012-Brief.pdf.
3. Tim Maurer, "Cyber Norm Emergence at the United Nations: An Analysis of the UN's Activities Regarding Cyber-security", Belfer Centre for Science and International Affairs, Harvard Kennedy School, Cambridge, September 2011, p. 22.
4. "Developments in the Field of Information and Telecommunications in the Context of International Security" Report of the Secretary General, United Nations, 5 August, 2005, p. 2, at http://www.un.org/disarmament/topics/informationsecurity/.
5. n. 3.
6. "Statement on Consensus Achieved by the UN Group of Governmental Experts on Cyber Issues", US Department of State, 7 June 2013, at http://www.State.gov/r/pa/prs/ps/2013/06/210418.htm. See also, the press release of the US State Department following the conclusion of the third GGE which reports considerable progress in narrowing differences.
7. UN General Assembly, A/RES/66/24, December 13, 2012.
8. Other than the permanent members, the other States included Argentina, Australia, Belarus, Canada, Egypt, Estonia, Germany, India, Indonesia and Japan. India has been a member of four out of five GGEs dealing with this issue, at http://www.un.org/ga/search/view_doc.asp?symbol=A/68/98.
9. The specific recommendations related to state responsibility were as follows:

   The group recommended that States cooperate to prevent harmful ICT practices and should not knowingly allow their territory to be used for internationally wrongful acts using ICT. It called for the increased exchange of information and assistance to prosecute terrorist and criminal use of ICTs. In doing so, the Group emphasized that States should guarantee full respect for human rights, including privacy and freedom of expression.

   A State should not conduct or knowingly support ICT activity that intentionally damages or otherwise impairs the use and operation of critical infrastructure.

   States should also take appropriate measures to protect their critical infrastructure from ICT threats.

   States should not harm the information systems of the authorized emergency response teams of another State or use those teams to engage in malicious international activity.

   States should encourage the responsible reporting of ICT vulnerabilities and take reasonable steps to ensure the integrity of the supply chain and prevent the proliferation of malicious ICT tools, techniques or harmful hidden functions.

   States have a primary responsibility to maintain a secure and peaceful ICT environment, international cooperation would benefit from the appropriate participation of the private sector, academia and civil society.

   States must not use proxies to commit internationally wrongful acts using ICTs, and should seek to ensure that their territory is not used by non-State actors to commit such acts.

   States must meet their international obligations regarding internationally wrongful acts attributable to them under international law. However, the indication that an ICT activity was launched or otherwise originates from the territory or the ICT infrastructure of a

State may be insufficient in itself to attribute the activity to that State. The Group noted that the accusations of organizing and implementing wrongful acts brought against States should be substantiated.

10. Ministry of Foreign Affairs, Cuba, Declaration by Miguel Rodríguez, Representative of Cuba, at the Final Session of the Group of Governmental Experts on Developments in the Field of Information and Telecommunications in the Context of International Security, *cubadiplomatica*, 23 June 2017, at misiones.minrex.gob.cu/en/un/Statements/71-unga-cuba-final-session-group-governmental-experts-developments-field-information.
11. Ibid.
12. Michele G. Markoff, "Explanation of Position at the Conclusion of the 2016–2017 UN Group of Governmental Experts (GGE) on Developments in the Field of Information and Telecommunications in the Context of International Security", US Department of State, 23 June 2017, at www.State.gov/s/cyberissues/releasesandremarks/272175.htm.
13. International Telecommunications Union, at http://www.itu.int/en/about/Pages/default.aspx.
14. n. 3, p. 29.
15. The WSIS was preceded by the Geneva Summit in 2003 where a plan of action was laid out under the rubric: WSIS Action Lines. Action Line 5 was with regard to cyber security. "Geneva Plan of Action." C5. Building Confidence and Security in the Use of ICTs, at groups.itu.int/stocktaking/About/WSISActionLines/C5.Cybersecurity.aspx.
16. "ITU Gateway for WSIS", at http://www.itu.int/itu-wsis/.
17. "Society, Internet", Internet Society Comment to the WCIT Preparations, 9 February 2012.
18. *ITU's Role with Regard to International Public Policy Issues Pertaining to the Internet and the Management of Internet Resources, Including Domain Names and Addresses* (Resolution 102 (Rev. Busan, 2014)), International Telecommunications Union, at www.itu.int/en/action/Internet/Documents/Resolution_102_pp14.pdf.
19. *Internet Protocol-based networks* (Resolution 101 (Rev. Busan, 2014)), International Telecommunications Union, at https://www.itu.int/en/action/Internet/Documents/Resolution_101_pp14.pdf.
20. *Strengthening the role of ITU in building confidence and security in the use of information and communication technologies* (Resolution 130 (Rev. Busan, 2014)), International Telecommunications Union, at https://www.itu.int/en/ITU-D/Cybersecurity/Documents/130revBusan.pdf.
21. Proposals for the Work of the Conference: ITU's role in Improving Network Functionalities for Evincing Trust and Confidence in IP based Telecom Networks, at http://www.itu.int/md/S14-PP-C-0098/en.
22. "Tunis Agenda for the Information Society", International Telecommunications Union, 18 November 2005.
23. "Resolution adopted by the General Assembly", United Nations, 2 February 2011, at http://www.un.org/ga/search/view_doc.asp?symbol=A/RES/65/141.
24. "Outcome Document of the High-level Meeting of the General Assembly on the Overall Review of the Implementation of the Outcomes of the World Summit on the Information Society", United Nations, 13 December 2015, p. 13.
25. The Internet Governance Forum (IGF), 24 June 2015, at www.intgovforum.org/cms/2015/IGF.24.06.2015.pdf.
26. Punneth Nagaraj and Aaarti Bhavana, "Multistakeholderism in action: Analyzing Indian Engagement at Global Internet Governance Institutions 2011–2015", Delhi 2016, pp.61–63.
27. "London hosts cyberspace security conference," *BBC*, 1 November 2011, at www.gov.uk/

government/news/london-conference-on-cyberspace-chairs-Statement.

28. Sterling, Bruce, "Cyberspace with Chinese Characteristics", *Wired*, Conde Nast, 8 October 2012, at www.wired.com/2012/10/cyberspace-with-chinese-characteristics-%E7%BD%91%E7%BB%9C%E7%A9%BA%E9%97%B4/.
29. "India to Host Global Conference on Cyberspace 2017: World's Largest Conference on Cyberspace," Press Information Bureau, 21 July 2017, at pib.nic.in/newsite/PrintRelease.aspx?relid=168850.
30. The full members are Russia, China, Kazakhstan, Kyrgyzstan, Uzbekistan, Tajikistan, India, and Pakistan; observers are Afghanistan, Belarus, Iran, and Mongolia; and dialogue partners are Armenia, Azerbaijan, Cambodia, Nepal, Sri Lanka, and Turkey.
31. "All States must play the same role in, and carry equal responsibility for, international governance of the Internet, its security, continuity and stability of operation, and its development in a way which promotes the establishment of multilateral, transparent and democratic international Internet governance mechanisms which ensure an equitable distribution of resources, facilitate access for all and ensure the stable and secure functioning of the Internet."
32. Letter dated 9 January 2015 from the Permanent Representatives of China, Kazakhstan, Kyrgyzstan, the Russian Federation, Tajikistan and Uzbekistan to the United Nations addressed to the Secretary-General, p. 5, United Nations, at http://daccess-dds-ny.un.org/doc/UNDOC/GEN/N15/014/02/PDF/N1501402.pdf.
33. Peter Wood, "China Conducts Anti-Terror Cyber Operations with SCO Partners", Jamestown Foundation, 19 October 2015, at jamestown.org/program/china-conducts-anti-terror-cyber-operations-with-sco-partners/. See also, "SCO Countries Hold Drill Targeting Cyber-Terrorism," English.news.cn, *Xinhua*, 6 December 2017, at www.xinhuanet.com/english/2017-12/06/c_136806108.htm.
34. Public Statement 2013 Plenary Meeting of the Wassenaar Arrangement on Export Controls for Conventional Arms and Dual-use Goods and Technologies, at http://www.wassenaar.org/publicdocuments/2013/WA%20Plenary%20Public%20Statement%202013.pdf.
35. Wassenaar Arrangement, at https://www.federalregister.gov/articles/2015/05/20/2015-11642/wassenaar-arrangement-2013-plenary-agreements-implementation-intrusion-and-surveillance-items.
36. "What Is the U.S. Doing About Wassenaar, and Why Do We Need to Fight It?", Electronic Frontier Foundation, 28 May 2015, at https://www.eff.org/deeplinks/2015/05/we-must-fight-proposed-us-wassenaar-implementation.
37. "Coalition of Security Companies Forms to Oppose Wassenaar Rules", Threatpost.com, at https://threatpost.com/coalition-of-security-companies-forms-to-oppose-wassenaar-rules/113794#sthash.XviOou8J.dpuf.
38. "Indian officials see cyber threats from Wassenaar Arrangement", *The Economic Times,* 19 June 2014, at http://articles.economictimes.indiatimes.com/2014-06-19/news/50711034_1_cyber-threats-inter-ministerial-panel-software-products.
39. Michael N. Schmitt, *Tallinn Manual on the International Law Applicable to Cyber Warfare,* prepared by the International Group of Experts at the Invitation of the NATO Cooperative Cyber Defence Centre of Excellence, Cambridge University Press, 2015.
40. Ibid., pp. 83–84.
41. Ibid., pp. 114–115.
42. Michael N. Schmitt, *Tallinn Manual 2.0 on the International Law Applicable to Cyber Operations*, Cambridge University Press, 2017, pp. 30–50.

43. "Rule No. 1 for Tech's Fight with China: Don't Talk about Your Fight with China", *Bloomberg*, 19 November 2014.
44. High-level Advisory Committee Established for World Internet Conference, World Internet Conference, Cyberspace Administration of China.
45. Adam Segal, "China's Internet Conference: Xi Jinping's Message to Washington", Net Politics, 16 December 2015, at http://blogs.cfr.org/cyber/2015/12/16/chinas-Internet-conference-xi-jinpings-message-to-washington/.

CHAPTER 4

Active Cyber Defence: An Analysis

Active Cyber Defence is a concept developed by researchers in the United States along the military as well as infosec tracks. As the name suggests, it distinguishes itself from passive cyber defence through more aggressive and robust actions that are conceivably in a grey area in terms of their legality. The Tallinn Manual makes the distinction between active and passive cyber defence as follows: The former is "a proactive measure for detecting or obtaining information as to a cyber intrusion, cyber attack, or impending cyber operation, or for determining the origin of an operation, that involves launching a pre-emptive, preventive or cyber counter-operation against the source". Passive cyber defence is defined as "a measure for detecting and mitigating cyber intrusions, and the effects of cyber attacks that does not involve launching a pre-emptive, preventive or cyber counter-operation against the source."[1]

Active Cyber Defence and Cyber Deterrence

Active Cyber Defence has been co-opted into the unfolding debate on cyber deterrence since it fits both into the deterrence by denial and deterrence by punishment paradigms. Cyber deterrence is sought to be achieved through the fusing of "intelligence, cyber defence, sanctions, diplomacy, and other policy tools together" to deter attacks.[2] Whilst both concepts have their own inherent problems when it comes to implementation, at present, in combination, they seem to offer the only framework for responding to cyber attacks. Even when carried out under state auspices, Active Cyber Defence has been problematic for apparent state sanction to break laws in the cause of national security; the logical conclusion that such sanctions should also be given to private entities is it's even more problematic aspect.

Active Cyber Defence was first conceptualised as an approach for the military to provide a more robust response than the existing passive defences such as antiviruses, firewalls, and intrusion detection systems, commonly in use as the first and only line of defence against cyber attacks.[3] The US Department of Defense formulated a definition in its Strategy for operations in Cyberspace published in 2011, which reads as follows:

> Active Cyber Defense is DoD's synchronized, real-time capability to discover, detect, analyze, and mitigate threats and vulnerabilities. It builds on traditional approaches to defending DoD networks and systems, supplementing best practices with new operating concepts. It operates at network speed by using sensors, software, and intelligence to detect and stop malicious activity before it can affect DoD networks and systems. As intrusions may not always be stopped at the network boundary, DoD will continue to operate and improve upon its advanced sensors to detect, discover, map, and mitigate malicious activity on DoD networks.[4]

As companies and organisations have been targeted by state-sponsored and non-state actors, they have also raised a demand to be allowed to incorporate "aggressive, external, offensive countermeasures", generically described as "hacking back" into the arsenal of responses available to them.[5] Their justification is that they are losing millions of dollars worth of intellectual property due to cyber espionage,[6] and governments lack the resources to prevent such theft.

Advanced Persistent Threats

Active Cyber Defence gained traction with the arrival of the Advanced Persistent Threats (APT), called so because threat actors remained undetected in networks for considerable periods of time. The working definition of an Advanced Persistent Threat as given by the National Institute of Standards and Technology (NIST) is as follows:

> An adversary [is one] that possesses sophisticated levels of expertise and significant resources which allow it to create opportunities to achieve its objectives by using multiple attack vectors (e.g., cyber, physical, and deception). These objectives typically include establishing and extending footholds within the information technology infrastructure of the targeted organizations for purposes of exfiltrating

> information, undermining or impeding critical aspects of a mission, program, or organization; or positioning itself to carry out these objectives in the future. The advanced persistent threat: (i) pursues its objectives repeatedly over an extended period of time; (ii) adapts to defenders' efforts to resist it; and (iii) is determined to maintain the level of interaction needed to execute its objectives.[7]

Advanced Persistent Threats were initially considered the sole forte of nation states, and hacker groups operating with nation state backing since they required a combination of resources and manpower that only states had the wherewithal to muster. Networks were compromised by exploiting vulnerabilities in software (called zero-days) yet to be discovered and patched by software companies.

The challenges posed by Advanced Persistent Threats are of a new order, with the attacker being able to easily surmount passive defences engineered to keep out automated worms and viruses. While the objective of viruses and worms was to compromise as many computers and networks as possible to pave the way for a variety of criminal activities, the main purpose behind the instigators of Advanced Persistent Threats is to intrude into a network, and stay for as long a period as possible without being detected. This is the case with both state sponsored actors after information as well as a new generation of criminal enterprises that find better returns from a more targeted than randomized activity.[8]

Among the earliest APT attacks were the so-called Moonlight Maze intrusions in 1998, with the attacker being able to easily surmount passive defences engineered to keep out automated worms and viruses. While the objective of viruses and worms were cyber espionage operations directed against government entities. The first known cyber attack APT operation against critical infrastructure was the Stuxnet attack on Iranian nuclear reactors in 2010. The first known operation against commercial entities was the so-called Operation Night Dragon, whose primary target was energy companies. In 2011, the information security company RSA's two-factor authentication system was compromised and used for subsequent APT attacks against US defence contractors. Many of these APT campaigns are global in nature and many government, private, and other networks have been affected by them.

Many of these attacks were believed to be carried out by proxy actors originating out of China and Russia. Though much of the Snowden revelations were focused on the intrusive surveillance of Western intelligence agencies, they also showed that there were units within Western intelligence agencies—such as the Equation Group tied to the National Security Agency (NSA)—that were carrying out similar activities.[9]

The Mechanics of Active Cyber Defence

The concept of Active Cyber Defence gained further traction with the kill-chain model that has been conceptualised to describe the various stages employed by APT Actors. The kill-chain model itself has been taken from military parlance to describe the various stages of a kinetic attack, and the differing opportunities for engagement and response. The stages in the military cycle are broken down into: i) find adversary targets suitable for engagement; ii) fix their location; track and observe; iii) target with suitable weapon or asset to create desired effects; iv) engage adversary; v) assess effects.

In a cyber attack scenario, an APT actor begins by:

- **Reconnaissance:** Research, identification and selection of targets, often represented as crawling Internet websites such as conference proceedings and mailing lists for email addresses, social relationships, or information on specific technologies.
- **Weaponization:** Coupling a Remote Access Trojan (RAT) with an exploit into a deliverable payload, typically by means of an automated tool (weaponizer). These are largely client application data files, such as Adobe Portable Document Format (PDF) or Microsoft Office documents.
- **Delivery:** Transmission of the weapon to the targeted environment, mainly through email attachments, websites, and USB removable media.
- **Exploitation:** After the weapon is delivered to the victim host, exploitation triggers the intruders' code. Most often, exploitation targets an application or operating system vulnerability; but it could also more simply exploit the users themselves, or leverage an operating system feature that auto-executes code.
- **Installation:** Installation of a remote access Trojan or backdoor on the victim system allows the adversary to maintain persistence inside the environment.

- **Command and Control** (C2): Typically, compromised hosts must beacon outbound to an Internet controller server to establish a C2 channel. APT malware especially requires manual interaction rather than conduct activity automatically. Once the C2 channel establishes, intruders have "hands on the keyboard" access inside the target environment.
- **Actions on Objectives**: Only now, after progressing through the first six phases, can intruders take actions to achieve their original objectives. Typically, this objective is data exfiltration which involves collecting, encrypting, and extracting information from the victim environment. Violations of data integrity or availability are potential objectives as well. Alternatively, the intruders may only desire access to the initial victim box for use as a hop point to compromise additional systems, and move laterally inside the network.[10]

Since each of these phases is dependent on the other, it only takes the disruption of any link in the chain to disrupt the whole process. Breaking an attack cycle into its constituent parts makes it, (i) easier to respond in each phase, and (ii) it reduces the inherent advantages enjoyed by the attacker by way of anonymity. The earlier the attacker is stopped in the cycle, the better are the chances of preventing and disrupting the attacks.

The responses to each segment of the kill-chain have been proposed as follows:

- **Reconnaissance**: mine and analyse open resources to provide indicators and warning of intrusion attempts;
- **Weaponize**: analyse artefacts to create high fidelity signatures to detect malicious activity;
- **Deliver**: understand the adversaries' tools and techniques for delivering messages to intercept them early;
- **Exploit**: leverage anti-exploitation and exploit detection techniques to find zero-day attempts;
- **Control**: employ robust intrusion detection signatures and tools to detect newly installed implants;
- **Execute**: instrument and configure internal networks to detect existing internal compromises;
- **Maintain**: deploy advanced host analysis to detect hidden implants and abnormal activity.[11]

The Kill-chain concept and response through the Active Cyber Defence method have enabled a more focused approach to deal with the threats from Advanced Persistent Threat actors, so much so that further analysis and elaboration on this methodology is an ongoing process. While the original model of Active Cyber Defence emphasised a technology-driven response, one view has it that a greater human element through the analysis of threat intelligence would increase the reliability of this model and make it less reactive. This would also necessarily include an element of sharing of information.[12] On the other hand, for the response to be in realtime there would be an increasing dependence on artificial intelligence and automated responses, raising further questions on command and control.[13]

Issues in Active Cyber Defence

Active Cyber Defence is a controversial concept, not least because much of the activities proposed are illegal under law, be it US or the laws of other countries.[14] At its most basic level, Active Cyber Defence is considered as undertaking four activities: local intelligence gathering; remote intelligence gathering; actively tracing the attacker; and actively attacking the attacker.[15] According to David Dittrich, excepting the first, the others are illegal[16] since they all undertake activity on remote networks. Such activities run the spectrum from benign to intermediate to aggressive. While benign refers to actions limited to own networks such as installing honeypots and scanning network traffic, intermediate refers to actions such as invasive tracebacks, and remote evidence collection. Aggressive responses veer on hacking back with actions such as remote exploitation, corruption of data, and denial of service attacks.[17]

The "rules of conduct" are still evolving, and vary depending on whether Active Cyber Defence is carried out under the auspices of the state or a private entity. Active Cyber Defence as a state action would necessarily fall under the rubric of countermeasures which, as yet, only exists as international customary law since the Draft Articles on the Responsibility of States for Internationally Wrongful Acts is still a work in progress in the UN system.

Since the objective of Active Cyber Defence is to mitigate the attack which might necessarily involve disabling the source of the attack, rules and norms governing the use of force such as proportionality, distinction, and necessity also have to be taken into account. Under the principle of necessity, forces must engage only in those actions necessary to achieve legitimate military objectives; and they must distinguish between lawful and unlawful targets such as civilians, and civilian property and innocent third parties. Proportionality referred to prohibition on excessive use of force.[18]

Difficulties with attribution further complicate the adaptation of these principles to cyberspace.

Relevant International laws include Article 2(4) of the UN Charter which prohibits the use of force,[19] Article 51 on the state's right to respond to an armed attack,[20] as well as customary international laws on non-intervention.[21] Typically, such laws only apply when there has been loss of life or irreparable damage to property. While the scope of Article 2(4) and Article 51 is only beginning to be debated with respect to cyber attacks, incipient responses such as NATO's declaration that attacks on member states would be taken up collectively under Article 4,[22] and a more recent declaration that a response would be formulated under Article 5[23] of the NATO Treaty.[24] The use of proxy actors and cyber "mercenaries" or APT actors is a way of evading state responsibility for actions in cyberspace.[25]

At the international level, even trans-border access to data for law enforcement purposes—a major feature of the Budapest Convention on Cybercrime—has proved to be controversial, and a major factor in the reluctance of states to accede to the Convention. Article 32 of the Convention on "Trans-border access to stored computer data with consent or where publicly available" states that:

> A Party may, without the authorisation of another Party:
>
> i) access publicly available (open source) stored computer data, regardless of where the data is located geographically; or
>
> ii) access or receive, through a computer system in its territory, stored computer data located in another Party, if the Party obtains the lawful and voluntary consent of the person who has the lawful authority to disclose the data to the Party through that computer system.

The explanatory report to the Convention notes that Trans-border access means "to unilaterally access computer data stored in another Party without seeking mutual assistance". Another caveat was that local laws and standards should be applied and followed while undertaking an investigation via the Budapest Convention.[26]

Private sector attempts to legitimise Active Cyber Defence have been bolstered by reports such as that of the Commission on the Theft of American Intellectual Property which recommends that "without damaging the intruder's own network, companies that experience cyber theft ought to be able to retrieve their electronic files or prevent the exploitation of their stolen information."[27]

For the private sector that has borne the brunt of APT attacks, Active Cyber Defence is seen as a new methodology that takes away the initiative from the attackers, and provides a more level playing field to the defenders. It reiterates many of the changes that have been called for by governments, including greater cooperation within the private sector since the attackers are using the same tools against different companies; moreover, companies would benefit greatly from sharing information on the attacks and how they responded in various stages of the kill-chain. As against rhetorical calls for greater cooperation, this gives a purpose to both private-private partnership as well as Public-Private Partnership. Such cooperation is also an effective response to research which shows that 90 per cent of attacks are discovered by third parties, including information security and antivirus companies as well as law enforcement agencies in those countries where they have sufficient expertise.

All this notwithstanding, remote intelligence gathering which is an integral part of Active Cyber Defence is tantamount to hacking and, therefore, illegal under prevailing National laws.[28] Moreover, Active Cyber Defence goes beyond intelligence gathering, and visualises more aggressive actions, including "retrieving data, shutting down systems, sabotaging data, infecting the attacker with malware, taking over the attacker's botnet, or hiring a botnet to attack the attacker."[29] In addition, as others have noted, information sharing unless through authorised channels—such as sector Information Sharing and Analysis Center (ISACS)—is also a violation of corporate laws. There is also the argument in the case of cyber criminals that such "vigilantism" only has short-term effects, and the cyber criminal can always re-build the infrastructure that has been disrupted.[30]

Many unintended consequences of such 'hack back' actions have been pointed out by analysts, ranging from the possibility of data in commandeered botnet computers being damaged through such actions, with the unintended consequences being amplified if these computers are being used for controlling critical infrastructure or belong to major corporations.[31]

All this notwithstanding, Active Cyber Defence also has many proponents, who see current models of response as inadequate to respond to current challenges. For instance, the Hudson Institute's report, *Cyber-Enabled Economic Warfare: an Evolving Challenge*, places this within a strategic context, noting that economic warfare is as old as human history, and has become even more potent when carried out through cyberspace."[32] Those advocating this proposal, when confronted with illegality under prevailing laws, tout justifications including self-defence, hot pursuit, and copyright; and, a new generation of information security companies are already undertaking such actions.[33] Infosec companies aver that their actions are misunderstood, and do not amount to hacking back or vigilantism; they are done with a view to increasing the cost to the attacker to the extent possible. Access to adversary networks would better enable attempts at attribution of attackers and their sponsors. The flexibility of actions to include deception, containment, and tying up adversary resources would also slow down exfiltration speeds, and increase the chances of mistakes. Some of the existing mechanisms—such as the sharing of Information which currently takes place through Information Sharing and Analysis Centres (ISACS)—would be upgraded so that information sharing takes place in realtime.[34]

Private companies point to instances of successful collaboration with law enforcement agencies, giving the example of the successful takedown of the Coreflood botnet in 2011. More recently, a number of cybersecurity firms performed "the first ever-private sponsored interdiction against a sophisticated state sponsored advanced threat group" code-named Axiom.[35] Approximately 43,000 computers were cleaned of malware believed to have been planted by this group, allegedly China based. The companies involved have promised that

> This is the beginning of what will hopefully be a long line of industry-coordinated efforts to expose these threat groups, and to do so without having to use law enforcement, to help corporations and governments around the world combat hackers.[36]

These arguments have found traction with the US government in a scenario where law enforcement agencies are unable to cope with the rash of cyber attacks.[37] Private companies have also found a favourable ear in the US Congress, with successive attempts at bringing in legislation to legalise Active Cyber Defence, the most recent being the discussion draft of the Active Cyber

Defense Certainty Act. This Act allows victims of cyber fraud to undertake ACD measures, including unauthorised access to other computers if it is for the purpose of retrieving data owned by them.[38]

The arguments for and against notwithstanding, the official position of the US authorities, reiterated in a speech by Assistant Attorney General Leslie R. Caldwell in May 2015, is that "freelance 'hacking back' and similar intentional intrusions onto third-party computers and networks can carry serious legal consequences and policy risks." She gave a number of reasons why hacking back could have negative repercussions all around. Firstly, hackback tactics posed a significant threat to innocent third parties whose infrastructure was used by malicious actors. Private hacking back would "needlessly expose such third parties, who often are unaware that their systems have been compromised, to intrusions, privacy violations, and potentially property damage." Secondly, hacking back could interfere with ongoing government investigations, and compromise evidence. Thirdly, hacking back carries the danger of dramatic escalation since the adversary could be anybody—from sophisticated cyber criminals to foreign intelligence services—who could retaliate even more viciously. Fourthly, companies might unwittingly break the laws of other countries in the course of their actions. Fifthly, such actions might be misperceived, and lead to international incidents. Sixthly, this did not provide a permanent solution to the problem and, at best, might provide only a temporary reprieve from APT actors.[39]

Active Cyber Defence in the Indian Context

Active Cyber Defence as a concept has not been much debated in the Indian context even though India has been a constant target of Advanced Persistent Threats as evidenced in a number of reports from the *Ghostnet Report* of 2009[40] to the *Shadows in the Cloud* Report in 2010,[41] followed by the *Operation Shady Rat* report in 2011,[42] and in the *Red October* and *Netraveller* Reports of 2013.[43] More recently, there were reports that hackers had broken into the server of the Airports Authority of India (AAI), and wiped data from an entire server in July 2014. A Pakistani cyber espionage campaign against Indian networks was highlighted in a report by Fireye in August 2014. Though researchers could not identify the specific victim organizations, they based their deductions on malware bundled with decoy documents relating to Indian issues. The malware sent data back to a US server to "make it seem like the attack originated from a US server."

What might come under the rubric of Active Cyber Defence has been attempted periodically, though very few instances are available in the public domain. In 2011, the National Technical Research Organisation, the intelligence agency tasked with cyber operations, was apparently behind the announcement of a Rs. 30,000 bounty to take over a botnet at a hacker's conference. The organisers announced that the contest had been successful, and the winning contestants had indeed hijacked the command and control servers of a Chinese botnet, though they had not gone beyond showing proof-of-concept.[44] However, subsequent news items indicated that the data on the C&C server had been removed, and was found to include sensitive files belonging to the Defence Research and Development Organisation.[45] Apparent government sponsorship of an illegal act was condemned by security researchers, both within the country and outside.[46]

Active Cyber Defence pre-supposes capabilities within the private sector and some element of technical expertise with law enforcement. Both are lacking in the Indian context. While India is a giant in information technology, there are comparatively few companies working on cybersecurity products and services. Even the few cybersecurity companies that there are, sometimes combine twin roles of producing their own products while also being vendors of foreign products. Ironically, many foreign cybersecurity companies have R&D facilities in the IT hubs of Bangalore, Pune, and Gurgaon.

Information sharing is a crucial element of Active Cyber Defence. A Joint Working Group (JWG) on engagement with the private sector on Cyber Security was established in July 2012, under the direction of the Deputy National Security Advisor. The JWG released a report in October 2012 detailing the guiding principles underpinning this exercise, and outlining a proposed roadmap for greater cooperation and coordination. With information sharing being crucial to combating cyber threats, the road map called for the establishment of Information Sharing and Analysis Centres (ISACs) in various sectors. ISACs established in critical sectors such as banking, telecom, and power are in various stages of development; but they are largely dependent on the nodal agencies/companies that have been identified in the various sectors. Unless the teething problems are identified and resolved, information sharing remains only a nominal activity. There is also very little in-house expertise in the law enforcement agencies which largely depend on technical agencies such as CERT-In.

As far as the legal aspects are concerned, Article 66 of The Indian IT Act of 2000 states in unambiguous terms that

(i) Whoever with the intent to cause or knowing that he is likely to cause wrongful loss or damage to the public or any person destroys or deletes or alters any information residing in a computer resource or diminishes its value or utility or affects it injuriously by any means, commits hack; and
(ii) Whoever commits hacking shall be punished with imprisonment up to three years, or with fine which may extend up to two lakh rupees, or with both.[47]

Other relevant sections are Section 85 which deals with the liability of companies to actions contravening the IT Act. The section reads as follows:

> Offences by companies: (i) Where a person committing a contravention of any of the provisions of this Act or of any rule, direction or order made thereunder is a company, every person who, at the time the contravention was committed, was in charge of, and was responsible to, the company for the conduct of business of the company as well as the company, shall be guilty of the contravention and shall be liable to be proceeded against and punished accordingly.

Given the ethical ambiguities and lack of clarity on the legality of Active Cyber Defence, Indian companies are, on record, not in favour of undertaking such activities. Under the circumstances, those companies that have been hacked and have sought to undertake actions that would come under the rubric of Active Cyber Defence (such as retrieving data) have been forced to employ companies on a cash basis. In general, Indian companies are more in favour of alternatives to Active Cyber Defence in order to combat cyber threats, in the form of adaptive cyber defence incorporating big data analytics, and the use of intelligent algorithms and big data to find patterns and responding to them. Active Cyber Defence is viewed as a peculiarly American formulation conceived to find justification for responding to attacks under American jurisprudence.

Active Cyber Defence is one among a spectrum of responses to the threats emanating from Advanced Persistent Threat actors. While many of its elements have proved to be controversial, especially with regard to "hacking back", it cannot be discounted entirely, especially in the face of recent successes. The moral of the story, as it were, is that threat actors are continuously finessing their capabilities and the responders, whether it is the government or the private sector, also have to do likewise. This should ideally be achieved through more cooperation between government agencies and the private sector, not through governments outsourcing their responsibility of being the overarching security provider to private companies or acquiescing to private sector demands that they take the lead in responding to APTs. Governmental agencies have to work on multiple fronts to discount such arguments. They have to acquire sufficient skills and capabilities to attain credibility to show that they can take the lead in cyber defence. They also have to work closely with the private sector in order to acquire the necessary threat intelligence, domain awareness, and technical expertise necessary to counter APT threats. There has to be better coordination within government when it comes to the sharing of information. The military will also have to be brought into the loop as cyber commands are set up, given the blurring of boundaries in cyberspace. The private sector will also have to be incentivised to create indigenous cybersecurity products as well as build up capabilities to produce threat intelligence.

NOTES

1. Michael Schmitt, Glossary, in the *Tallinn Manual on the International Law Applicable to Cyber Warfare,* Cambridge: Cambridge University Press, 2013, p. 257.
2. "Russian Hacking: the Real Threat Lies Ahead", *The Christian Science Monitor,* 9 May 2017, at www.csmonitor.com/USA/Politics/2017/0509/Russian-hacking-the-real-threat-lies-ahead.
3. See, for instance, Tiong Pern Wong, "Active Cyber Defense: Enhancing National Cyber Defense", Master's thesis, the Naval Postgraduate School, December 2011.
4. Pentagon, Department of Defense Strategy for Operations in Cyberspace, July 2011, p. 7, at www.defense.gov/news/d20110714cyber.pdf.
5. Robert Dewar, "The 'Triptych of Cyber Security': A Classification of Active Cyber Defence", Proceedings of 6th International Conference on Cyber Conflict, Estonia, Tallinn, 2014, p. 10.
6. Josh R. Rogin, "NSA Chief: 'Cybercrime Constitutes the "greatest Transfer of Wealth in History'," *Foreign Policy,* 9 July 2012, Web, 21 November 2014, at http://foreignpolicy.com/2012/07/09/nsa-chief-cybercrime-constitutes-the-greatest-transfer-of-wealth-in-history/.
7. Appendix B, National Institute of Standards and Technology, Information Security, March 2011, p. B-1, at http://csrc.nist.gov/publications/nistpubs/800-39/SP800-39-final.pdf.
8. See, for instance, paper presented by Elie Bursztein et al., *Handcrafted Fraud and Extortion: Manual Account Hijacking in the Wild,* 2014 Internet Measurement Conference, Vancouver Canada: November 5–7, 2014.
9. Thomas Fox-Brewster, "Equation = NSA? Researchers Uncloak Huge 'American Cyber Arsenal'", *Forbes,* February 18, 2015, accessed 19 November 2015, at http://www.forbes.com/sites/thomasbrewster/2015/02/16/nsa-equation-cyber-tool-treasure-chest/ In this regard, there have also been writings that describe the intrusive surveillance carried out by the NSA as a manifestation of active cyber defence. See, for instance, Keir Giles

and Kim Hartmaan, "Socio-Political Effects of Active Cyber Defence Measures," proceedings of the 6th International Conference on Cyber Conflict, Estonia, Tallinn, 2014, Web, 19 November 2015, p. 24.

10. E.M. Hutchins, M.J. Cloppert and R.M. Amin, "Intelligence-Driven Computer Network Defence Informed by Analysis of Adversary Campaigns and Intrusion Kill Chains," Proceedings of the 6th International Conference on Information Warfare and Security (ICIW 11), Academic Conferences Ltd., 2010, p. 4, at http://www.lockheedmartin.com/content/dam/lockheed/data/corporate/documents/LMWhite-Paper-Intel-Driven-Defense.pdf.
11. Mitre Corporation, *Active Defense Strategy for Cyber,* July 2012, at https://www.mitre.org/sites/default/files/publications/active_defense_strategy.pdf.
12. Matt Hartley, "Strengthening the Cyber Kill Chain with Cyber Threat Intelligence", Insight partners Blog, 8 September 2014, at http://www.isightpartners.com/2014/09/strenghtening-cyber-kill-chain-cyber-threat-intelligence-part-1-of-2/.
13. See, Catriona Heinl, "Artificial (Intelligent) Agents and Active Cyber Defence: Policy Implications", Paper presented at the 6th International Conference on Cyber Conflict, Tallinn, Estonia, June 2014.
14. "The Computer Fraud and Abuse Act in the USA", at https://www.law.cornell.edu/uscode/text/18/1030.
15. Adam Segal, "Hacking Back, Signalling, and State-Society Relations", Blog post, 1 March 2013, at http://www.cyberdialogue.ca/2013/03/hacking-back-signaling-and-state-society-relations-by-adam-segal/.
16. Kenneth Himma, and David Dittrich, "Active Response to Computer Intrusions." *The Handbook of Information Security*, 2005, Web 15 September 2014, at http://papers.ssrn.com/sol3/papers.cfm?abstract_id=790585.
17. David Dittrich, Active Response to Computer Intrusions, in the *Handbook of Information Security* (ed.), Hossein Bidgoli, Hoboken, N.J.: John Wiley & Sons, Inc., 2005, p. 3.
18. Jody R. Westby, "Cyber War vs. Cyber Stability", in R. Ragaini (ed.) International Seminar on Nuclear War and Planetary Emergencies: 42nd Session, 2010, World Scientific Publishing Co., p.100.
19. Article 2(4) of the U.N. Charter states that member states "shall refrain in their international relations from the threat or use of force against the territorial integrity or political independence of any state, or in any other manner inconsistent with the Purposes of the United Nations; see, UN Charter, Chapter 1, art. 2, paragraph 4, at http://www.un.org/en/documents/charter/chapter1.shtml.
20. Article 51 states that "Nothing in the present Charter shall impair the inherent right of individual or collective self-defence if an armed attack occurs against a Member of the United Nations...", UN Charter, Chapter XVII, at http://www.un.org/en/documents/charter/chapter7.shtml.
21. Oona A. Hathaway, "The Drawbacks and Dangers of Active Defence", Paper presented at the 6th International Conference on Cyber Conflict, Tallinn, Estonia, June 2014, p. 39.
22. Article 4 of the Treaty states that "The Parties will consult together whenever, in the opinion of any of them, the territorial integrity, political independence or security of any of the Parties is threatened", NATO Treaty, at http://www.nato.int/cps/en/natolive/official_texts_17120.htm.
23. Article 5 of the Treaty states that "the Parties agree that an armed attack against one or more of them in Europe or North America shall be considered an attack against them all and consequently they agree that, if such an armed attack occurs, each of them, in exercise of the right of individual or collective self-defence recognised by Article 51 of the Charter

of the United Nations, will assist the Party or Parties so attacked by taking forthwith, individually and in concert with the other Parties, such action as it deems necessary, including the use of armed force, to restore and maintain the security of the North Atlantic area," NATO Treaty, at http://www.nato.int/cps/en/natolive/official_texts_17120.htm.

24. Steve Ranger, "NATO Updates Cyber Defence Policy as Digital Attacks Become a Standard Part of Conflict", *ZDNet*, 30 June 2014, Web 15 November 2014, at http://www.zdnet.com/nato-updates-cyber-defence-policy-as-digital-attacks-become-a-standard-part-of-conflict-7000031064/.
25. James Farwell and Rafal Rohozins, "*The New Reality of Cyber War*", 22 October 2012, Web, at http://www.defenceiq.com/cyber-defence/articles/the-new-reality-of-cyber-wa/.
26. "Transborder access to data", Council of Europe, Cybercrime Convention Committee, Guidance Note # 3, Article 32, 5 November 2013.
27. Commission on the Theft of American Intellectual Property, National Bureau of Asian Research, 2013, p. 6.
28. These include the Computer Fraud and Abuse Act (CFAA) in the USA, the Computer Misuse Act of 1990 in the United Kingdom, and the Information Technology Act 2000 (as amended by the IT Act, 2008).
29. Jody Westby, "Caution: Active Response to Cyber Attacks Has High Risk", *Forbes* Magazine, 29 November 2012, at http://www.forbes.com/sites/jodywestby/2012/11/29/caution-active-response-to-cyber-attacks-has-high-risk/.
30. Becca Lipman, "Hacking the Hackers: The Legal Risks of Taking Matters into Private Hands", *Wall Street & Technology*, 23 September 2014, at http://www.wallstreetandtech.com/security/hacking-the-hackers-the-legal-risks-of-taking-matters-into-private-hands/a/d-id/1315980?_mc=RSS_WST_EDT.
31. CSIS/DOJ Active Cyber Defense Experts Roundtable, March 10, 2015, US Department of Justice, Web 19 December 2015, at http://www.justice.gov/sites/default/files/criminal-ccips/legacy/2015/05/18/CSIS%20Roundtable%205-18-15.pdf.
32. Samantha Ravich (ed.), *Cyber-Enabled Economic Warfare: An Evolving Challenge*, Hudson Institute, August 2015, Web 21 November 2015, at http://www.hudson.org/research/11408-cyber-enabled-economic-warfare-an-evolving-challenge. See especially, Juan Zavarte, "A New Cyber-Privateering Framework", a chapter within the report, which is also published separately in *The Convergence of Financial and Cyber Warfare and the Need for a 21st Century National Security Response*", by the Foundation for Defence of Democracies, July 2015, Web, 21 November 2015, at http://www.defenddemocracy.org/media-hit/juan-zarate-the-cyber-financial-wars-on-the-horizon/.
33. Dmitri Alperovitch, *Active Defense: Time for a New Security Strategy*, 25 February 2013, Web, at http://www.crowdstrike.com/blog/active-defense-time-new-security-strategy/.
34. Ibid.
35. Ellen Nakashima, "Researchers Identify Sophisticated Chinese Cyberespionage Group", The *Washington Post*, 28 October 2014, Web, at http://www.washingtonpost.com/world/national-security/researchers-identify-sophisticated-chinese-cyberespionage-group/2014/10/27/de30bc9a-5e00-11e4-8b9e-2ccdac31a031_story.html. The executive summary of the report on the Axiom takedown can be found at http://www.novetta.com/files/9714/1446/8199/Executive_Summary-Final_1.pdf.
36. Ibid.
37. "Would the U.S. Really Crack Down on Companies That Hack Back?", *Bloomberg*, 30 December 2014, at www.bloomberg.com/news/2014-12-30/why-would-the-u-s-crack-down-on-companies-that-hack-back-.html.
38. United States, Congress, House – Judiciary Committee, "H.R. 4036 - Active Cyber Defense

Certainty Act." *H.R. 4036 - Active Cyber Defense Certainty Act,* 10 Dec. 2017. 115th Congress, 1st session, bill H.R. 4036, www.congress.gov/bill/115th-congress/house-bill/ 4036 https://tomgraves.house.gov/uploadedfiles/discussion_draft_active_cyber_ defense_ certainty_act_2.0_rep._tom_graves_ga-14.pdf.

39. Leslie Caldwell, "Assistant Attorney General Leslie R. Caldwell Delivers Remarks," Georgetown Cybersecurity Law Institute, Washington, May 2015, Department of Justice, Web, 19 Dec. 2015, at http://www.justice.gov/opa/speech/assistant-attorney-general-leslie-r-caldwell-delivers-remarks-georgetown-cybersecurity.
40. "Tracking GhostNet: Investigating a Cyber Espionage Network," *Tracking GhostNet: Investigating a Cyber Espionage Network,* 1 Sept. 2009,athttp://www.infowar-monitor.net/ 2009/09/tracking-ghostnet-investigating-a-cyber-espionage-network/, accessed on 02 June 2010.
41. *Shadows in the Cloud: Investigating Cyber Espionage,* 2.0 Joint Report, Toronto, Information Warfare Monitor and Shadowserver Foundation, 2010.
42. Dmitri Alpevorich, *Revealed: Operation Shady Rat: An investigation of targeted intrusions into 70+ global companies, governments and non-profit organizations during the last 5 years,* 2011, at http://www.mcafee.com/us/resources/white-papers/wp-operation-shady-rat.pdf, accessed on 15 January 2012.
43. *The "Red October" Campaign: An Advanced Cyber Espionage Network Targeting Diplomatic and Government Agencies,* Kaspersky, 2012, at http://www.securelist.com/en/analysis/ 204792262/Red_October_Diplomatic_Cyber_Attacks_Investigation,accessed on 23 October 2013.
44. "Hacking Govt Server Fetches Him Rs 35k", *Goacom,* 14 March 2013, at http:// www.goacom.com/goa-news-highlights/7139-hacking-govt-server-fetches-him-rs-35k.
45. "Chinese 'hack' DRDO Computers; Antony Seeks Report", The *New Indian Express,* 14 March 2013, at http://www.newindianexpress.com/nation/article1500336.ece.
46. David Dittrich, "The Honeynet Project", *A New Infosec Era? Or a New Infosec Error?,* 11 March 2013, at http://www.honeynet.org/node/1031.
47. Information Technology Act, 2000, Gazette of India, p. 19, at http://www.dot.gov.in/sites/ default/files/itbill2000_0.pdf. The term "hacking" was subsequently removed in the IT Amendment Act, 2008, supposedly because of objections from institutes that taught ethical hacking courses.

CHAPTER 5

Critical Information Infrastructure Protection: National Practices and Perspectives

Physical infrastructures, such as electricity generation, transmission and distribution, transportation services of railways and airways, telecommunications, and services in the form of banking, healthcare, or taxation are the underpinnings of modern society. Their seamless operations and availability are essential not just for social and economic growth, but for national security as well. Moreover, these infrastructures are heavily dependent upon information technology and information systems for a variety of technical and management functions, such as operations, controls, maintenance, and communications. Information Technology (IT) infrastructure as a backbone of information systems in every sector or industry enables the efficient storage, processing, and transmission of data or information. Also, the control systems in the manufacturing industry, chemical processing, or petroleum refining plants, electricity generation, or transmission installations, etc. utilize programmable logical controllers and computers for smooth, reliable, and continuous operations.

In general, every infrastructure is dependent on other infrastructures for its core functions: for instance, banking depends upon the telecommunication network for connectivity among branches. The day to day requisites of water, electricity, transportation, fuel, and food supply chains, banking and financial services, communications, etc. are completely dependent on each other, giving rise to dependence and interdependence among themselves. The interactions are sometimes complex, and hard to simulate or comprehend as they are spread across organizational boundaries or physical and political borders.

Interdependence could also induce vulnerabilities, and a minor disruption at a point or in one of the infrastructures may have a rippling effect across multiple critical infrastructures,[1] which could be debilitating or disruptive. For instance, a disruption in electricity supply to the railway or metro systems can bring such essential services to a standstill.

The definitions, scope, and protection practices for Critical Infrastructure (CI), and their subset of Critical Information Infrastructure (CII), represent different national perspectives. However, the concerns regarding the protection of CI and CII are being raised equally at the international and multilateral platforms. In its Resolution 58/199, the United Nations General Assembly recognised the complexity of the network of critical information infrastructure components, exposing them to a growing number and a wider variety of threats and vulnerabilities that raise new security concerns.[2] The 2015 UNGGE report also underscored the issue of attacks targeted against critical infrastructure, and associated information systems.[3] The G8 countries had adopted the "Principles for Protecting Critical Information Infrastructures",[4] and the Organisation for Economic Co-operation and Development had also recommended a policy framework for the development of national policies and international cooperation.

As a multilateral platform for government officials, The Meridian Process aims to facilitate an extensive exchange of ideas and cooperation among governmental bodies on issues relating to the protection of CII. The European Union Agency for Network and Information Security (ENISA), established in 2004, coordinates European Union (EU) wide efforts to secure and protect the availability of Information and Communication Systems that are essential for the operation of CI.[5] The EU model is applicable to other regions, where infrastructures such as the electricity grids, civil aviation, railways, navigation services, and energy supply chains are closely knit among countries. Today, Critical Infrastructure Protection (CIP) and Critical Information Infrastructure Protection (CIIP) have been acknowledged as vital components of national security policy. Governments have, therefore, adopted stern policy measures which have also led to the establishment of new organisations with clear mandates to devise and execute comprehensive strategies encompassing multiple stakeholders; the industry, academia, the private sector, and government entities.

National responses and strategies differ in their respective capacity, approach, and implementations. There is no definitive strategy, but a number of factors shape these strategies, such as the perspectives from system level technicalities, business perspectives, law-enforcement perspectives and, above all, the national security perspective. Policy initiatives are also subject to the priorities set by the governments under their respective domestic circumstances, such as legal and regulatory frameworks, relationships between the public and private sector, governments' commitments, etc.

CIP in the USA

The first step towards CI Protection in the USA was a Presidential Decision Directive on Critical Infrastructure Protection (PDD-63) in 1998.[6] It was enforced in retrospect of the increasing reliance of the US economy upon interdependent and cyber-supported infrastructures, and the threat of non-traditional attacks on the infrastructure and information systems. Over the last two decades, the Office of the President of the USA has been closely associated with policy making for CI Protection, apparently as one of the top national security priorities.

Five years after PDD-63, in the wake of the terrorist attack on the World Trade Centre in September 2001, the Homeland Security Presidential Directive 7 (in 2003) established a national policy for Federal departments and agencies to identify and prioritize critical infrastructure and to protect them from terrorist attacks.[7] By then, the USA deemed terrorist attacks against its critical infrastructure to be plausible enough to threaten its national security, cause mass casualties, weaken the economy, and damage public morale and confidence. This directive superseded the 1998 directive, and authorised an integrated National Plan for Critical Infrastructure and Key Resources Protection. A decade later, in 2013, the Presidential Policy Directive-21 (PPD-21) further refined and clarified the critical infrastructure-related functions, roles, and responsibilities across the Federal Government.[8] More integrated with the new developments, the directive called for a revision in the National Infrastructure Protection Plan (NIPP), addressing the implementation of the new directive; the requirements of amended Homeland Security Act; and alignment with the National Preparedness Goal mandated by Presidential Policy Directive / PPD-8 for National Preparedness.

The new Trump administration released a Presidential Executive Order on Strengthening the Cybersecurity of Federal Networks and Critical Infrastructure in May 2017.[9] It encompasses the management of cybersecurity risks, and seeks reports, recommendations and inputs from various Federal agencies and departments on a wide cross-section of areas varying, from the implementation of PPD-21 requisites to the strategic options for deterrence in cyberspace.

The Department of Homeland Security (DHS) is the apex body for CIP implementation in the USA, and promotes a national unity of effort and coordinates the overall federal effort for the security and resilience of critical infrastructure. The latest framework for government and private sector participants in the critical infrastructure community to work together to manage risks and achieve resilience, the NIPP, was published in 2013.[10] The initial version of NIPP was released in 2006, and it was further revised in 2009. Over the years, the NIPP has evolved, adapting and streamlining the current risks, policy imperatives, and the strategic environment, envisioning secure and resilient physical and cyber critical infrastructure.

NIPP has been developed through a collaborative process involving stakeholders from all the sixteen critical infrastructure sectors, all fifty states, and from all levels of government and industry.[11] It also implements the requirements of PPD-21, imbibing the unique characteristics, operating procedures, and risk profiles of each of the critical infrastructure sectors. Sector-Specific Agency (SSA), designated for each of the sixteen critical infrastructure sectors, is tasked with development and implementation of a sector-specific plan, applying the NIPP concepts to the unique characteristics and conditions of the specific sectors.[12] A key policy document, NIPP strengthens the partnerships among owners and operators; Federal, State, local, tribal, and territorial governments; regional entities; non-profit organizations; and academia.

The Sector Coordinating Councils are basically self-organized, self-run, and self-governed private sector councils which facilitate discussions on strategies, policies, activities, and issues. They draw on representation from the owners and operators of the industries within the critical infrastructure sectors.[13] Cross-Sector Councils coordinate cross-sector issues, initiatives, and interdependencies, and are composed of the chairs and vice-chairs of the Sector Coordinating Councils.[14] Government Coordinating Councils enable interagency, inter-governmental, and cross-jurisdictional coordination within and across sectors. They partner with Sector Coordinating Councils, and consist of representatives from across various levels of government.[15] Figure 5.1 depicts the interactions among the Government Coordinating Council, the Sector Coordinating Council, and Cross Sector Councils for the critical infrastructure sectors.

Figure 5.1: Critical Infrastructure Protection Apparatus in the US

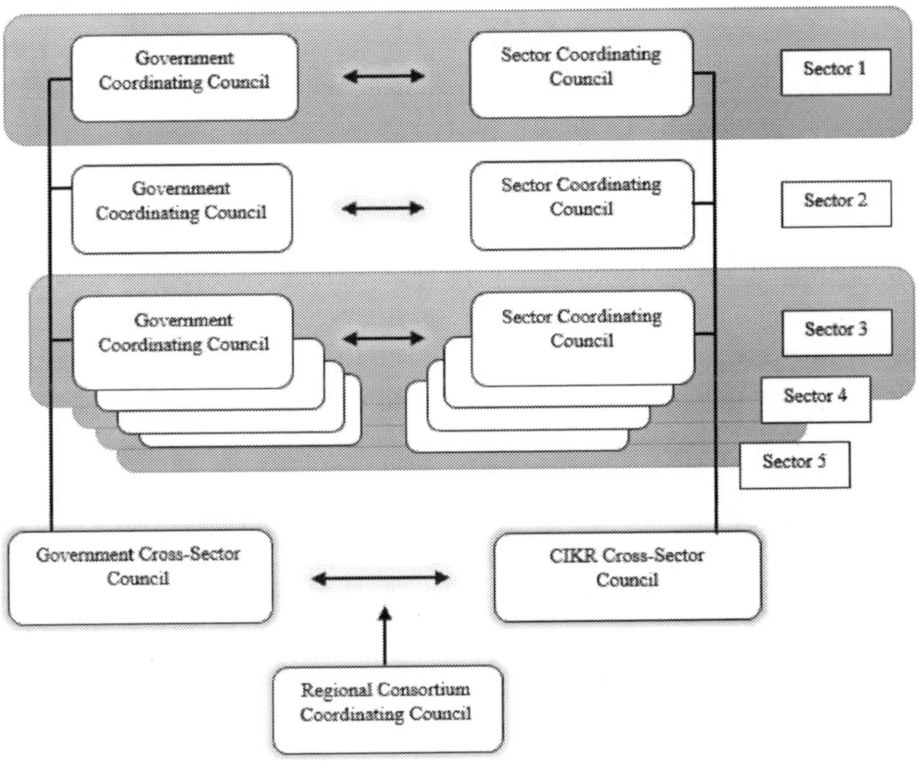

Source: US Department of Homeland Security.

The Federal Senior Leadership Council, comprises senior officials from Sector-Specific Agencies and other Federal departments and agencies, State, Local, Tribal, and Territorial; the Government Coordinating Council, consists of representatives from across SLTT government entities; the Regional Consortium Coordinating Council, comprises regional groups and coalitions, integrating efforts, expertise, interests and representation of all the partners in national critical infrastructure security and resilience.

The USA has identified 16 critical infrastructure sectors, whose assets, systems, and networks are considered so vital that their incapacitation or destruction would have a debilitating effect on security, national economic security, national public health, and safety.[16] CIP in the USA has come a long way, particularly after the September 11, 2001 attacks. Given the strategic imperatives and the implications for national security and the economy, etc., the successive administrations have kept a close tab on the developments and implementations of the policies since 1998. In the present shape, definite roles and responsibilities are assigned to Federal agencies, key departments, the intelligence community, and the other stakeholders; they ensure representation and reciprocate their requirements. However, throughout the last two decades, the President's office has exercised its authority to direct policy measures and supervise the implementation of CIP policies. The Policy Directives and Executive Orders seek time bound inputs or reports from the agencies. This also underscores the quantum of significance critical infrastructure protection holds for the USA.

CIP in the United Kingdom

The Cabinet Office of the Government of the UK oversees the implementation of policy initiatives under the strategic framework for the resilience of critical infrastructure. As a strategic step, the National Cyber Security Centre (NCSC) was set up in October 2016 as the UK's apex body on cybersecurity, and is a part of the Government Communications Headquarters (GCHQ).[17] The NCSC works in close coordination with other government departments, law enforcement agencies, the defence establishment, intelligence and security agencies, and international partners. The NCSC was set up to strengthen cybersecurity for the UK, as cyber threats are reaffirmed as being one of the most significant risks to UK interests in the National Security Strategy of 2015. The NCSC has absorbed and replaced the CESG (the information security arm of GCHQ), the Centre for Cyber Assessment (CCA), the Computer Emergency Response Team UK (CERT UK) and the cyber-related responsibilities of the Centre for the Protection of National Infrastructure (CPNI). The CPNI provides security advice on physical and personnel security. Its cybersecurity/information assurance responsibilities have been absorbed into the NCSC.

The CPNI follows the "Protective Security" methodology, building security measures or protocols in the design itself, to deter, detect, or minimize the consequences of an attack.[18] The CPNI works in close coordination with the other institutions specializing in security and counter-terrorism: the National Counter Terrorism Security Office (NaCTSO); the Counter Terrorism Security Advisor (CTSA) network, and the recently established NCSC.[19] Respective sectors of the critical infrastructure are led by government departments for devising and implementing protective security measures. The CIP practice in the UK is government driven, but the inputs from private sector and the academia are ratified, and are an integral part of "Protective Security". Protective security encapsulates national threat perception, and is based upon an 'all-risks' model encompassing a wide section of risks emanating from terrorism, espionage threats, and natural hazards.

One of the distinct approaches in the UK strategy for CIP is the categorisation of infrastructure according to its "criticality", and impact assessment using a criticality scale. Such an exercise ensures critical elements receive the utmost priority. This criticality scale assesses impacts of an adverse event or an attack on: a) the delivery of the nation's essential services; b) the economy, arising from the loss of essential services; and c) human life.[20] The details of criticality and impact assessment are laid out in the sector resilience plan, which has evolved over five revisions since 2010.[21] The sector resilience plan, produced annually, evaluates the relevant risks identified in the National Risk Assessment.

The UK has one of the more mature CIP practices. It integrates national efforts under the NCSC, drawing in synergies among the expertise and experience residing with the governmental departments and the specialised agencies from the intelligence community, the computer emergency response teams, and also from the CPNI. Sector resilience plans also have a scientific approach to the identification of criticalities and interdependencies within the infrastructures. This gives the government, and the owners and operators of critical infrastructure, a clear understanding of the interdependencies or complex interactions, to then prioritise the most critical of all the elements.

CIP in Australia

The *Critical Infrastructure Resilience Strategy* of the Australian Government, the primary policy statement, charts out the plan for the practical implementation of policy. The strategy aims to strengthen the resilience of critical infrastructure, ensuring their continued operations in the face of a wide spectrum of adverse events, primarily by adopting an "all-hazards" approach.[22] Two of the major terrorist attacks—the one on 11 September 2001 in New York and the Bali bombings in 2002—compelled the Government to establish a national Critical Infrastructure Strategy for Australia.[23]

The Attorney-General's Department is the lead agency for this task, responsible for both the Critical Infrastructure Centre and the Trusted Information Sharing Network (TISN). The Critical Infrastructure Centre, established recently in January 2017 under the Department of Home Affairs, coordinates the management of national security risks to critical infrastructure on four priority high-risk sectors: telecommunications, electricity, water, and ports.[24] The centre consolidates the expertise and capability across the government for the efficient management of national security risks, working in close consultation with state and territory governments, regulators, and private owners and operators. The centre also integrates the expertise of industry and state and territory governments, who are the primary operators, owners, and regulators of critical infrastructure, with the proficiency of intelligence agencies in security threats and vulnerabilities. The government has a consultative approach. A discussion paper was released in February 2017,[25] seeking inputs from the respective stakeholders on the better management of proposed functions of the centre, modalities, and methodologies for the Critical Infrastructure Asset Register.

Recognising the enduring threat from terrorist attacks, the Australia-New Zealand Counter-Terrorism Committee published national guidelines in 2015. The guidelines deem the governments to play an active role in this regard. However, the owners and operators of CI are also held responsible for addressing the security of their respective assets and the continuity of their core business functions.[26] The guidelines also mention identifying criticality levels (low-significant-major-vital) and mapping interdependencies for prioritising CI in terms of their criticality from a national perspective. The Critical Infrastructure Program for Modelling and Analysis is central to the practice of modelling and simulating the dependency relationships of the systems part of critical infrastructure.

Risk and Resilience are the two cornerstones of strategy, broadly in a non-regulatory business-government partnership model.[27] It warrants the owners/

operators of CI to respond to the risks from multifaceted social, economic, technological, and environmental changes. It includes natural disasters, pandemics, negligence, accidents, criminal activity, computer network attack, and terrorism. The Trusted Information Sharing Network (TISN) and the Critical Infrastructure Advisory Council (CIAC) are the two primary mechanisms to build a partnership between the government and business sector.[28]

The Government of Australia established the TISN in 2003 and, since then, it has been the prime mechanism for engagement, business-government information sharing, and resilience building initiatives.[29] The TISN also facilitates an important informal link between the industry sectors and their respective regulatory agencies from the sectors of aviation, communications, offshore oil and gas, and banking.[30] The state and territory governments are also key participants in the TISN.

Figure 5.2: TISN Governance Structure

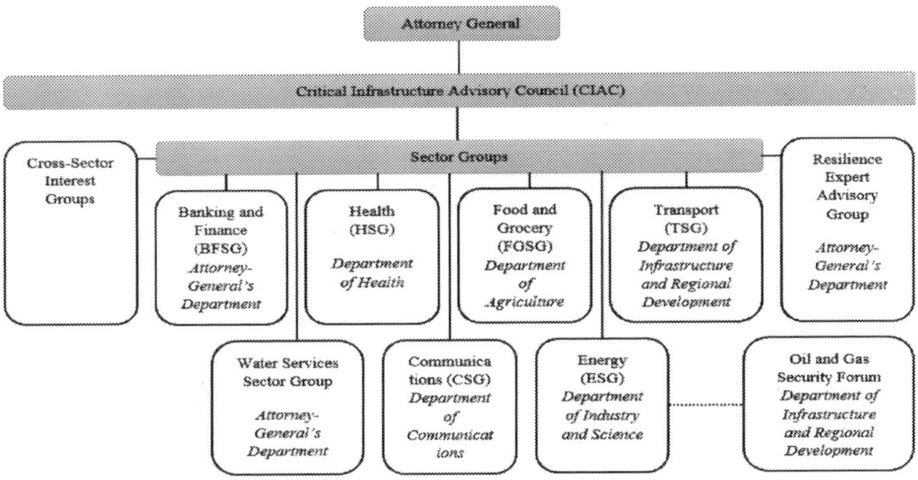

Source: Australian Government, TISN.

The Critical Infrastructure Advisory Council (CIAC), consisting of the Chairs of each of the TISN Groups, senior Australian Government representatives, and senior State and Territory government representatives, provides coordination and strategic guidance.[31] The Industry Consultation on National Security (ICONS) facilitates business-government engagement on national security matters, between business leaders and the Attorney-General.[32] In addition, Cross-Sectoral Interest Groups assist in exploring solutions for the issues cross-cutting different sectors. The Resilience Expert Advisory Group, with representation from state and territory governments, critical infrastructure owners and operators as well as academia and research organisations, promotes the concept of organisational resilience within the business community of critical infrastructure sectors.[33]

The Australian strategy for CIP has the right blend of governmental control and guidance at one end, and a conducive environment for a business-government partnership at the other. Information sharing, through both formal and informal networks/links, between businesses/private players and state/territory government is a key enabler. The TISN and non-regulatory business-government partnerships are the core of the national CIP strategy. *Australia's Critical Infrastructure Resilience Strategy* lays down four principal outcomes with definite action points. It will be further reviewed in 2020, for both progress evaluation and the assimilation of relevant changes. Identifying the key elements of critical infrastructure systems and their dependencies as part of the strategy using a methodical approach assists the owners and operators in prioritising efforts and measures.

CIP in China

China's Cyber Security Law of 2016 (in Article 31) has defined the national critical information infrastructure as

> the information facilities that are related to national security, national economy and people's livelihood, which have been damaged, destroyed or lost, may seriously endanger the national security and public interests, including but not limited to the provision of public communication, radio and television transmission network, energy, finance, transportation, education, scientific research, water conservation, industrial manufacturing, health care, social security, public utilities and other areas of important information systems, and important Internet applications.[34]

The Cyber Security Law sees CII protection as a common responsibility of the government, enterprises, and the society as a whole by using a combination of technology and management practices, and simultaneously strategies for protection and deterrence.[35] The Cyber Security Law came into effect on 01 June 2017.

The Cyber Security Law places strong emphasis on the protection of critical information infrastructure, holding the operators responsible for evaluating their respective cybersecurity risks and other potential risks annually.[36] Taking a step forward, pursuant to Article 31 of the Cyber Security Law of China,[37] the Cyberspace Administration of China (CAC) published a Draft Regulation on the Protection of Critical Information Infrastructure in July 2017, soliciting public opinion and comments.[38] At one end, the draft regulation clarifies the scope of CII and elaborates on the roles and responsibilities of the operators for network protection; at the other end, it sets out obligations—and even penalties—for the operators if they fail to do so.

According to the Draft Regulation, the scope of critical information infrastructure protection extends to the following entities: (i) government organs and units in the industries or fields of energy, finance, transportation, water conservancy, health, education, social security, environmental protection, and public utilities; (ii) information networks such as telecommunications networks, radio and television networks, and the Internet; and units providing cloud computing, big data, and other large scale public information network services; (iii) scientific research and production units in fields such as the national defence, large equipment industry, chemical industry, and food and medicine; (iv) news units such as radio stations, television stations, and news services; and (v) other key units.

The Draft Regulation obligates the operators to procure key network equipment and network security products that meet relevant national standards. Such products and services also need to undergo a network security review. The Draft Regulation also provisions fines if the operators fail to perform the desired security obligations, or violate the provisions, despite orders and warnings. The fines are extendable to both the operator and the responsible persons in charge, and might also lead to suspension of the relevant business or the revocation of business licenses.

China has also published an International Cyberspace Cooperation Strategy in March 2017. The strategy underscores China's interests in the security of information infrastructure, and it has a section on "Global Information Infrastructure Development and Protection".[39] It also accentuates China's willingness to shape the discourse at the international diplomacy front.

Legal measures in the form of a Cyber Security Law and Draft Regulations on the Protection of CII are China's first steps in the direction of securing and protecting CII. These steps, along with China's proposed Strategy for International Cyberspace Cooperation, underscore the importance China's government now enshrines in cybersecurity and the security of its CII. However, information pertaining to the developments, technical and policy measures, assessment of these measures, etc. is inadequate to form an analysis. CIP in China is at a nascent stage, and the steps, as of now, signify the Chinese government's commitment and resolve to address these issues. Nevertheless, the Chinese government has underscored the pertinent role of private industries and enterprises in this endeavour; and is deriving cues from the collaborative approach between the government and the private sector as it is practiced across the globe.

CIP in India

In pursuit of global efforts to protect CI and CII, India has also accentuated domestic efforts in recent years. The legal framework to address threats emanating from cyber terrorism to the CII took shape in the form of the IT Act, 2008. Section 66F of the IT Act identifies cyber terrorism to be a threat to CII as it could be used to "threaten the unity, integrity, security or sovereignty of India or to strike terror in the people or any section of the people."[40] In accordance with Section 70A of the Act, the National Critical Information Infrastructure Protection Centre (NCIIPC), a national nodal agency responsible for all the measures relating to the protection of CII, was established under the auspices of the National Technical Research Organization in January 2014.[41]

The NCIIPC aims to reduce the vulnerabilities of the CII against threats emanating primarily from cyber terrorism and cyber warfare. The roles and responsibilities of the NCIIPC are widespread, and vary from providing strategic leadership and coherence across government to coordinate, share, monitor, collect, analyze, and forecast national level threats to the CII for policy guidance, expertise sharing, and situational awareness.[42] As a part of its mandate, the NCIIPC issues regular guidelines, advisories, and vulnerability or audit notes to the operators of CII. It holds frequent consultations with stakeholders, including the private sector, and works in close coordination with the Indian Computer Emergency Response Team (CERT-In).[43] It has been tasked with generating awareness among public and private enterprises, as well as sensitising senior management from the CII operators. As a standardisation attempt, the NCIIPC has published framework cybersecurity evaluation, control guidelines, and Standard Operating Procedures for Auditing/Incident reporting,[44] to ensure that the requisite security mechanisms are built into the CII as key design features.[45]

The NCIIPC draws its Advisory Committee from the Ministry of Home Affairs; the Ministry of Law & Justice; the Department of Telecommunications, the Department of Electronics & IT; the Ministry of Defence, CERT-In, the National Security Council Secretariat, and the Cabinet Secretariat of the Government of India. It has representation from the Intelligence Bureau as well as Industry and State Governments. The five principal stakeholders are the CII owners/operators; service providers to the CII; the NCIIPC; CERT-In; and law enforcement agencies.

Within a short span of time, NCIIPC has held close consultations with the key government entities, strategic and public enterprises as well as the operators/owners from the private sector. It has been working incessantly towards the identification of "Protected Systems", whose security thereafter is in the ambit of the NCIIPC. Protected Systems then have increased policy and technology measures for information security in place, such as periodical Vulnerability/Threat/Risk assessment, cyber crisis management plan, information security audits, and logs analysis for networking and communication devices, systems and services. It has also begun to conduct security audits for identified critical sectors, and even rolled out a draft cybersecurity manual specific to the controls and requirements of the power sector.[46] Power & Energy and Banking & Financial Services sectors have apparently emerged as priority areas for the government. NCIIPC is spearheading initiatives like cyber audits of key public sector banks and workshops and training programs for the leadership and human resources from these sectors.

India's CIIP practices are gradually maturing towards a collaborative framework, under the auspices of a specialised agency to facilitate the technical and management concerns of the operators. The NCIIPC has a fair representation of all the stakeholders. Nevertheless, CIIP is an evolving and dynamic process, subjective to the specific conditions, legal frameworks, and regulatory environments of the state. Therefore, the best of the models, practices, and technologies are available for exchange, which could be adapted and reflected upon in respective domestic policy initiatives.

Securing India's CII: Learning from Global Practices

The practice of designating Sector-Specific Agencies (in the USA) or lead departments (in the UK) seeks to address the problems of specific sectors, based on their unique characteristics and attributes. As of now, India has designated sectoral CERTs in the Power Sector only.[47] The same model could be extended to other sectors at the earliest, with a government department as the lead. Also, self-organized, self-run, and self-governed private sector councils facilitate intense discussions, and ensure the representation of owners and operators in the policy making process. The private sector in India should be encouraged and facilitated for organising such councils to reap their true benefits. Most important, cross-sector coordination is vital for India to resolve issues and interdependencies which practically cut across different sectors, but find limited platforms. Critical Infrastructure Cross-Sector Council in the case of USA or similar practices in the UK or Australia are the prime examples for India to discern cross-sector understanding of the systems, networks, assets, and dependencies involved. This may further be augmented with a detailed and scientific analysis, based on strong methodology and an understanding of cross sector dependencies. In addition, a qualitative or quantitative methodology to identify key resources within the CI or CII would be quite beneficial to prioritise protection efforts for the CI and CII. These scientific methods to determine criticality would certainly aid policy makers in India in the effective implementation of policy measures and streamline coordination across the board. Speaking under the Chatham House rule at a workshop on the "Geopolitics of Cyberspace: Creating Space for India" held at IDSA, one of the experts pressed the importance of the Meridian Process, which currently is the only platform where national level policy experts gather to discuss critical information infrastructure protection. The speaker underscored the anticipated benefits for India as a part of Meridian Process, with access to global best management and policy practices.

In addition to the existing apparatus for national security matters, government-private sector engagement on the pertinent issues of national security could be established in India—similar to ICONS in Australia. This can help the government place national security issues up for resolution before the private sector, which is an essential component of India's CII and, going forward, is certainly going to play a prominent part. The IT Act mandates the CII to appoint a Chief Information Security Officer (CISO), who directly liaisons with the NCIIPC for all technical or administrative measures. In response to the survey question on the role of CISOs in India, the CISOs from some of the key power sector entities at the workshop on "Critical Information Infrastructure: Securing the Power Sector", (held at IDSA) converged on the point that, for CISOs to be productive and effective, they should have direct access to the board and a core team composed of executives from all the departments, using IT to deliver their key functions. Cybersecurity, in essence, should not be the sole responsibility of the IT department thereof; rather it should be a team work with active involvement of all major departments, whether it is human resources, finance, administration, legal, safety, or operations.

The CISOs at the workshop unanimously agreed that NCIIPC and CERT-In should also involve private sector entities in policy making practices proactively, and periodically review or interact with the organisation's board. Such efforts can help the board of these organisations to realize the importance of CII protection, particularly in the wider context of national security. As nodal and premier agencies mandated for cybersecurity in India, they should also promote collaboration between industry and academia for Research and Development in this domain, and aid their capacity or skill building. Most important, in the CII sectors, NCIIPC should make disclosures of vulnerabilities mandatory.

In line with the emerging practice across the globe, backed by legal measures, the government may also mandate a thorough review and testing of all imported products sourced to be integrated or deployed in the critical sectors. It should also build a domain- specific security mandate for the sectors, given their unique characteristics, attributes, and requirements. The CISOs also called for regular interaction with apex bodies, between the sectors, and among peer organisations for sharing the best of their technical and management practices pertaining to cybersecurity.

The deliberations at the workshop related to the expectations of the CISOs from the government, affirmed the view that the government should perform the dual role of supporter as well as regulator for cybersecurity practices. However, the government alone cannot achieve the objectives of CII protection.

Therefore, Public-Private Partnership (PPP) mechanisms should assist the operators and owners of critical sectors in audits, execute training programs, and establish test beds or labs to augment the requisite capacity for testing and certification. They can also innovate with training programs, and make them specialized and periodic. Such thought out and well-designed training programs for technical personnel (probably using simulations or demonstrations), auditors, and users (on awareness and basic cyber hygiene education) would be quite productive and effective. The responsibility of protecting CII does not begin or end with the government. In fact, all the entities—be they the government, the private sector, or the academia—are responsible in enabling the desired response. They all have to complement each other's capabilities and leverage them effectively for a concerted effort towards protecting India's Critical Information Infrastructure.

NOTES

1. Steven M. Rinaldi, James P. Peerenboom, and Terrence K. Kelly, "Identifying, Understanding and Analyzing Critical Infrastructure Dependency", *IEEE Control Systems Magazine* (USA), p. 14, v. 21, n. 6, 2001, pp. 11-25.
2. United Nations General Assembly, "UNGA resol: 58/199: Creation of a global culture of cyber security and the protection of critical information infrastructures", A/RES/58/199, 30 January 2004.
3. "Group of Governmental Experts on Developments in the Field of Information and Telecommunications in the Context of International Security", United Nations General Assembly, 22 July 2015, p. 6.
4. G8 Principles for Protecting Critical Information Infrastructures (Adopted by the G8 Justice & Interior Ministers), May 2003, p. 1.
5. "Critical Information Infrastructure", European Union Agency for Network and Information Security, at https://www.enisa.europa.eu/topics/critical-information-infrastructures-and-services/cii?tab=details, accessed on 18 September 2017.
6. "PDD-63: Critical Infrastructure Protection", *Clinton Digital Library*, 20 May 1998, at https://clinton.presidentiallibraries.us/items/show/12762, accessed on 19 September 2017.
7. US Department of Homeland Security, "Homeland Security Presidential Directive 7", 17 December 2003, at https://www.dhs.gov/homeland-security-presidential-directive-7, accessed on 19 September 2017.
8. Office of the Press Secretary - The White House, "Presidential Policy Directive - Critical Infrastructure Security and Resilience", February 12, 2013, at https://obamawhitehouse.archives.gov/the-press-office/2013/02/12/presidential-policy-directive-critical-infrastructure-security-and-resil, accessed on 19 September 2017.
9. "Presidential Executive Order on Strengthening the Cybersecurity of Federal Networks and Critical Infrastructure", Office of the Press Secretary, The White House, 11 May 2017, at https://www.whitehouse.gov/the-press-office/2017/05/11/presidential-executive-order-strengthening-cybersecurity-federal, accessed on 18 September 2017.
10. "National Infrastructure Protection Plan", US Department of Homeland Security, at https://www.dhs.gov/national-infrastructure-protection-plan, accessed on 19 September 2017.

11. Ibid.
12. "Sector-Specific Agencies", US Department of Homeland Security, at https://www.dhs.gov/sector-specific-agencies, accessed on 19 September 2017.
13. U.S. Department of Homeland Security, "National Infrastructure Protection Plan - 2013", pp. 10-12.
14. Ibid.
15. Ibid.
16. U.S. Department of Homeland Security, "Critical Infrastructure Sectors", at https://www.dhs.gov/critical-infrastructure-sectors, accessed on 19 September 2017.
17. "About us", National Cyber Security Centre, at https://www.ncsc.gov.uk/about-us, accessed on 22 September 2017.
18. "About CPNI", Centre for the Protection of National Infrastructure, at http://www.cpni.gov.uk/about/, accessed on 22 September 2017.
19. "Who we Work With", Centre for the Protection of National Infrastructure, at http://www.cpni.gov.uk/about/Who-we-work-with/, accessed on 22 September 2017.
20. "Critical National Infrastructure", Centre for the Protection of National Infrastructure, at http://www.cpni.gov.uk/about/cni/, accessed on 22 September 2017.
21. "Sector Resilience Plans", Cabined Office of the Government of UK, at https://www.gov.uk/government/collections/sector-resilience-plans, accessed on 22 September 2017.
22. "Critical Infrastructure Resilience", Attorney-General's Department: Australian Government, at https://www.ag.gov.au/NationalSecurity/InfrastructureResilience/Pages/default.aspx, accessed on 22 September 2017.
23. "Critical Infrastructure Resilience Strategy: Policy Statement", Attorney-General's Department: Australian Government, 2015, p. 1.
24. n. 22.
25. Australian Government, "Critical Infrastructure Centre, Strengthening the National Security of Australia's Critical Infrastructure: A Discussion Paper".
26. "National Guidelines for Protecting Critical Infrastructure from Terrorism", Australia-New Zealand Counter-Terrorism Committee, 2015, p. 3.
27. Australian Government, "Critical Infrastructure Resilience Strategy: Plan", 2015, p. 1.
28. "Critical Infrastructure Resilience Strategy", Australian Government, 2010, p. 4.
29. "Trusted Information Sharing Network for Critical Infrastructure Resilience", at http://www.tisn.gov.au/Pages/default.aspx, accessed on 25 September 2017.
30. n. 26, p. 4.
31. n. 26, p. 2.
32. n. 26, p. 3.
33. "Organisational Resilience", Australian Government, at https://www.organisationalresilience.gov.au/partners/Pages/default.aspx, accessed on 25 September 2017.
34. "National Cyberspace Security Strategy: Full Text", People.com, 27 December 2016, at http://politics.people.com.cn/n1/2016/1227/c1001-28980829.html, accessed on 25 September 2017, translated to English using Google translator.
35. Ibid.
36. "Overview of China's Cybersecurity Law", KPMG China, February 2017, p. 11.
37. "2016 Cybersecurity Law", China Law Translate, 07 November 2016, at http://www.chinalawtranslate.com/cybersecuritylaw/?lang=en, accessed on 25 September 2017.
38. "Circular of the State Internet Information Office on the Public Opinion on the Protection of Critical Information Infrastructure Security Regulations (Draft for Soliciting Opinions)", National Internet Information Office, 10 July 2017, at http://www.cac.gov.cn/2017-07/

11/c_1121294220.htm, accessed on 25 September 2017, translated to English using Google translator.

39. "Full text: International Strategy of Cooperation on Cyberspace", *Xinhua*, 01 March 2017, at http://news.xinhuanet.com/english/china/2017-03/01/c_136094371_5.htm, accessed on 25 September 2017.
40. "The Information technology Act, 2008", Ministry of Law, Justice and Company Affairs, Government of India, 2008, p. 25.
41. "About Us", National Critical Information Infrastructure Protection Centre, at https://nciipc.gov.in, accessed on 27 September 2017.
42. Ibid.
43. Ibid.
44. Ibid.
45. "Guidelines for Protection of Critical Information Infrastructure", National Critical Information Infrastructure Protection Centre, 16 January 2015, Version 2.0, p. 1.
46. Released at the India Smart Grid Week, organised by India Smart Grid Forum, in February 2016.
47. "Power Sector: Information Sharing and Analysis Centre", Central Electricity Authority, at http://www.cea.nic.in/isacpower.html, accessed on 27 September 2017.

CHAPTER 6

India's Technology Challenges: Encryption, Quantum Computing and Artificial Intelligence

Technology is the key driver of the advances made in the information age, whether it is lightning fast communications, the exponential rise in computing power and data storage capacity, keeping data safe from unauthorized access or disclosure, or even deriving intelligence from large data sets. Open markets, *laissez-faire* economic systems, and innovation ecosystems have made technology march ahead of laws and regulations. The foremost challenge for developing nations is not just technology absorption but also protecting sensitive information, ensuring privacy for citizens, securing technology supply chains, and inculcating ecosystems which enable innovation. While computing, information and communication technologies accentuate economic growth, social development and connectivity, governments continue to grapple with their security, regulatory, ethical, and legal dimensions.

The Debates on Encryption

The security of data and information is a global concern, spanning governments, industry, armed forces, and academia. Mathematicians, information scientists and engineers are finding novel cryptographic techniques, methods, and algorithms to secure information and data using encryption. Cryptography is extensively used to secure banking and financial transactions, access to personal information, and secure secrets and/or communications in the armed forces, governments, and commercial organisations. For information security, cryptography-based technologies use a multitude of security functions such as encryption algorithms, message digest functions, Hashed Message Authentication Code (HMAC) functions, secret key exchange algorithms, and digital signatures. The technologies used to establish authentication, secure digital payments, protect the confidentiality of information, and defend the ownership of intellectual property are all based on the science of cryptography. Encryption algorithms, as an application of cryptography, perform this technologically intensive task to secure data and information not just during transmission but also in storage, and probably will do so for decades in the future. The last decade has witnessed an exponential rise in the growth and use of technology for encryption. It is being used by businesses that wish to secure their trade secrets, and consumers who want to prevent any unauthorized access to their credit card details and, very recently, even to their text chats.

Cryptography has been an ancient technique, dating back to 1900 BC, and has been fundamentally used to protect secrets. In modern history, the wars of the 20th century witnessed military applications of cryptography, with the German-made Enigma, the British Type X, and the American SIGABA being prominently used during World War II. Given the strategic imperatives and military applications, both the armed forces and governments have designed and developed cryptographic applications and machines in utmost secrecy. Encryption technology and products have been treated as dual-use items, and have been subject to strict export control rules and regulations. Throughout the Cold War, diplomatic efforts and export control regimes—led by the West—ensured that encryption products did not fall into the hands of adversaries.

Cryptography was eventually brought into the public domain during the 1970s. Academic research, international publications, peer reviews, the standardization of encryption algorithms, and the influx of software engineering led to open discussions, debates, and vast advancements in the applications of cryptography, for both civilian and military uses. Ever since then, the development of cryptography and encryption as disciplines of both mathematics and engineering, have been done under the close scrutiny of governments and their security agencies. Privacy advocates and civil society have also kept a close vigil on these advancements, and reacted sharply to the development of backdoors[1] in the standards and algorithms.

The present day debate on Encryption is just another iteration of the surveillance and encryption debate of the 1970s, which rose again in the 1990s.

The discussion in the 1970s was centred on the key length of the encryption algorithms, which were allegedly weakened deliberately. The debate in the 1990s surrounded the question of building encryption in personal devices which, later on, was settled in favour of stronger encryption methods.

Today's debates, akin to those in the past, have arisen out of the friction between the legitimate requisites of law enforcement and the rights of the populace for privacy under the respective constitutions of different nation states. Particularly after the Snowden revelations, there has been a sudden rush to encrypt personal information. But this haste for encryption, mostly on part of the private entities, has spawned a policy conflict. It is essentially the conflict between the legitimate applications of encryption to secure information from unauthorised access, and the possibilities of its interference with the investigations related to crime and law enforcement. The tussle between the government and technologists is basically over the deliberate attempts either to limit the strength of encryption keys and/or algorithms, or to enable lawful access to encrypted data and communications.

The First Crypto War

The long running policy debate over encryption—better known as the Crypto Wars—actually began in the 1970s, with conflicting ideas over whether technology enterprises (such as IBM and Digital Equipment Corporation) could export hardware and software built with strong encryption, and whether academicians could freely publish their research in cryptography. The debate continued throughout the 1980s over whether the National Institute of Standards and Technology (NIST) or the National Security Agency (NSA) should control the development of standards for encryption algorithms. Even during the 1990s, the US government used export control mechanisms which sought to prevent private enterprises such as Microsoft and Netscape from using strong encryption algorithms in their products, such as web browsers. The onset of the Internet, however, brought an end or a pause to the Crypto wars. It also made encryption a commodity for the use of the common man.

In the 1990s, three broad solutions to address the policy conundrum of encryption were promulgated. The first option under consideration was to adopt a relatively weak cryptography, which could be broken whenever required to do so. The second choice was to adopt very strong cryptography which would not allow wiretapping. The third proposed option was Clipper, a technology that allowed very strong cryptography protecting communications and files against unauthorized access, but could also enable law enforcement agencies to do a wiretap. The Clipper Chip, a microchip inserted into consumer hardware telephones, was supposed to provide strong cryptographic tools to the citizens without undermining the ability of law enforcement and intelligence agencies to access decrypted communications.[2] The proposal faced an immediate backlash from technical experts, privacy advocates, and industry leaders, who were concerned about the security and economic impact of the technology, in addition to the obvious concerns over civil liberties.[3]

Other governments also followed suit, with proposals for encryption licensing that would require copies of encryption keys to be held in escrow maintained by trusted third parties. This idea of "software key escrow" also kept floating throughout the 1990s, but privacy, security, and economic concerns outweighed the potential benefits. With the Internet boom, this idea faded out slowly, and the option of stronger cryptography prevailed.[4] After 2000, the US government also removed the restrictions on the sale of strong encryption, basically to reap the benefits of Internet as it moved towards commercialization and a value creator for the economy. The first crypto war was mainly about deliberately weakening encryption algorithms or their implementation as well as stringent export controls to limit the access of this technology to a few trusted nations.

Encryption Algorithms: A History of Weakening the Standards

By virtue of being a signals intelligence agency, the NSA has a vast surveillance network; it also owns one of the world's most advanced cryptanalysis[5] infrastructure. The NSA is known to be closely engaged with the process of developing encryption standards, along with the National Institute of Standards and Technology (NIST). For a long time, this engagement has given rise to scepticism regarding the very process of standardization, and some of the technical changes in the algorithms carried out at the behest of the NSA. It began with the development of DES in the 1970s, primarily to address the requirements of the government of USA for data security. The IBM developed the cipher LUCIFER,[6] which apparently underwent modifications after technical consultations with NSA[7]—such as reducing the key size from 128 bits to 56 bits,[8] and the mysterious S-boxes, alleged to be a "backdoor".[9]

Random numbers are vital to most of the modern day cryptographic applications such as generating session keys or public keys, authentication, as well as nonce for digital signatures, etc. An insecure random number generator can compromise the security of an entire cryptographic system such as the RSA, which is one of the widely used public key encryption standards. In 2007, Bruce Schneier had raised suspicion on the NIST standard "Dual Elliptic Curve Deterministic Random Bit Generator (Dual_EC_DRBG)", to probably be a "backdoor".[10] RSA heavily relied on this standard. In response to the public concern on the trustworthiness of Dual_EC_DRBG, NIST removed it from the draft guidance on random number generators in 2014.[11] Perhaps, the episode of weakening standards meant for international usage has tarnished NIST's reputation as a trustworthy government entity in the global endeavour for information security.

Cryptanalysis is essentially driven by geopolitical and security compulsions, and is carried out both against the adversarial nation states as well as against friendly nations. The NSA has over time influenced encryption standards covertly to retain access to any information it perceives to be important for national security. News reports and independent analysts suggest that NSA has been spending heavily to build backdoors[12] and eavesdrop on internet traffic, possibly facilitated by web service providers and networking equipment manufacturers,[13] such as Cisco and Juniper networks.[14] NSA is also suspected of influencing the industry standard of encryption by blocking the publication of a number of academic papers on encryption. In the USA, since early 2016, this debate became public over the government's attempts to influence, pierce, and degrade commercial technology for information security in its quest to protect common citizens from acts of terrorism.[15] The government has also tried other ways to maintain access to the data and information stored on millions of personal devices, both through legitimate and illegitimate means.

The Debate on Lawful Access to Encrypted Data and Communications

In early 2008, the US Federal Bureau of Investigation (FBI) used the term "going dark". It was meant for criminals shrouding their communications using encryption. These concerns, however, were confined to the closed circles of policy and law makers in the USA. In a watershed moment, the Snowden revelations not only spilled this discourse into the public domain but also exposed the massive Internet surveillance program run by US law enforcement and intelligence agencies.

After losing the noted public battle (to insert 'back doors') in the 1990s, the NSA is alleged to have run a vast anti-encryption operation called "Bullrun."[16] The operation involved building superfast computers and software to break encryption codes, and foster partnerships with key US and foreign technology firms in a bid to imbed access points into their security or technology products. Taking a serious note of this, the global community and foreign governments furiously expressed their discontent. The surveillance network covering friendly nations also triggered uproar. This also fuelled the desire and need for strong encryption. As a reflex, Google began encrypting all Gmail data that flowed between its data centres in March 2014; Yahoo also followed suit within a few weeks. In another path breaking move in 2014, Apple also rolled out Full Disk Encryption in the next version of its operation system iOS 8. Following suit, Android also made Full Disk Encryption as a default feature rather than an optional one.

Quite soon, the encryption debate acquired an international colour. The then UK Prime Minister, David Cameron argued for a ban on messaging services without a decryption capability, while promoting his government's planned surveillance bill. Amidst the heated debate, WhatsApp deployed end-to-end encryption in its messaging application for its 1 billion plus users across the world in April 2016.

As a technology, encryption was developed to protect data and information from unauthorized access, safeguard the identity of the users and enhance privacy. It also enabled secure banking transactions, the privacy of communications, and the security of stored data for consumers. Eventually, coming out of the closets of the armed forces, governmental agencies, and mathematicians, encryption emerged as an important policy issue, with a bearing on crime, civil rights and liberties, national defence, and economic competitiveness.[17] Under unprecedented public scrutiny and concerns over mass surveillance, companies are gradually deploying stronger encryption solutions even on personal devices such as smartphones and tablets. This completely forbids even their own access to data stored on these personal devices, or the ability to decrypt it.

Encryption has become a part and parcel of everyday life. It protects financial and banking transactions, private conversations, and a wide array of online activities as discrete as a flight bookings or online shopping. Beyond that, encryption also has a dark side. The ease of access to encryption products and their usability has led to the proliferation of encryption technology in grey and black markets. Criminal syndicates, sellers and buyers of illicit goods traded at the "Silk Road" online black market in the Dark Web use Onion Routing, which is meant for anonymous communication. Cyber criminals are using strong encryption technologies to develop ransomware variants, which are increasingly being used to extort money from the victims. Wannacry ransomware, which created havoc across 150 countries in May 2017,[18] is also a fallout of advancing encryption technology being put into adverse use.

Law enforcement and security agencies have unequivocally raised their concerns over this proliferation of encryption as a detriment to their investigations in the acts of crime and terrorism. Stronger unbreakable encryption deployed by messaging apps or by the mobile manufacturers for hard drives prevents access to messages, communication details, and other content, which could probably be useful for investigation or prosecution purposes. Both as a technological reason and part of their commitment to users' privacy, manufacturers and service providers themselves do not have access to the content on the devices. This has triggered an intensive debate on the possibilities of lawful access to encrypted data, as and when the requirement arises. The debate appears to be a zero sum game between the requirements of security and law enforcement establishments and the privacy of the individual, with different versions appearing in different countries. Nation states are trying to find the right balance between security and privacy, based on the existing laws and regulations related to privacy, governance structures and due consideration to relevant practices and social factors.

Lawful Access to Encrypted Data: The Global Scenario

The deployment of encryption on smartphone and other personal devices is estimated to increase the global share of end-to-end protected traffic from 18 percent to around 22 percent by 2019.[19] Encryption has varied applications and interpretations. Individual needs for privacy and anonymity are way different from enterprise encryption meant for securing communications, data, and intellectual property. Governments and armed forces have massive classified information which needs utmost protection against unauthorised access and espionage. As a security threat, end-to-end encryption, particularly in messaging apps, enables terrorists and criminals to evade surveillance. For instance, Telegram supports Perfect Forward Secrecy in its messaging app,[20] in order to keep past communications safe. This seriously undermines investigations in the cases of crime and terrorism. Owing to threats from a multitude of actors, regulating the use and development of encryption products and services for law enforcement and national security purposes is quite common across the globe. These practices even extend to the use of backdoors and key escrows, or weakening encryption standards.

There is no global consensus on the pertinent issue of "lawful access to encrypted communications" owing to the divergent views of major players. At one end, in the USA, the UK, and few European countries, public voices in the favour of privacy and civil liberties are very prominent. These curtail the government's ability to restrict strong encryption algorithms. Russia and China, on the other end, have their own surveillance mechanisms to keep a tab on the personal online activities of individuals through encryption licensing mechanisms and collusion with service providers. In some countries, laws require individuals to comply with the investigations, and provide access to the plain text. The research, development, usage, and commerce of cryptographic products is also strongly controlled by the state through licensing systems.

The Bureau of Industry and Security of the US Department of Commerce administers and regulates export controls and the licensing policy of the government of USA.[21] The USA has one of most stringent licensing and export control regimes to restrict the exports of dual-use technology, including strong encryption, to countries where it runs the risk of being put into adverse use. In the USA, however, there is presently no law for key disclosures; but one of landmark cases related to decryption of data stored on electronic devices has been fought in the US judicial system. The faceoff between Apple and the FBI in the San Bernardino case of 2016 attracted global attention: the FBI compelled Apple to decrypt the phone of a dead terrorist, invoking the All Writs Act.

The USA also houses extensive technical expertise on the subject, and globally renowned cryptographers like Ronald L. Rivest, Peter G. Neuman, Whitfield Diffie and Bruce Schneier have overtly supported stronger encryption, often criticizing the government's policies and laws. A recent study report of The National Academy of Sciences, authored by well-respected cryptographers and technologists, lawyers, members of law enforcement, and representatives from the industry, expounds on the risks and modalities of exceptional access to the plaintext of encrypted communications and stored data.[22] It expands on the four possibilities of either taking no legislative action; providing additional resources to access plaintext; and devising legislation for either device vendors and service providers to grant access or to develop a technical approach.[23] The tussle over encryption has not left US legislators untouched, whether it is the California phone decryption bill or the draft bill authored by Senators Diane Feinstein and Richard Burr (termed as Compliance with Court Orders Act of 2016) mandating compliance with any authorized court order for data, or to render it "intelligible".[24]

In April 2017, China's State Cryptography Administration published a draft Encryption Law for public comment. It covered various aspects such as scientific research, production, sale, import and export, testing, certification, use, and regulation and administration of cryptography under the three broader categories: "common" encryption, "core" encryption, and "commercial" encryption.[25] It makes provisions for telecommunication operators and Internet service providers to provide "decryption technology support" to public security organs and national security organs.[26] China also implements a licensing system for commercial cryptography products, and Article 16 of the proposed law controls the import and export of cryptography.[27] The regulatory regime for cryptography in China is undergoing an overhaul. The government is abolishing approval requirements for the manufacturing, sale, and use of commercial encryption products,[28] shifting the focus of regulation from supply chain to finished encryption products. Most importantly, in order to ensure unrestricted access to encrypted communications in the cases of terrorism, China's Anti-Terrorism Law also requires telecommunication operators and Internet service providers to provide technical interfaces, decryption, and other technical support and assistance to public security organs and state security organs.[29]

In the case of Russia, the Federal Law on Licensing Certain Types of Activity covers general licensing procedures related to the dissemination, development, and production of encryption (cryptographic) facilities.[30] The Federal Security Service (FSB) is the licensing authority. Article 15 obligates public authorities, enterprises, institutions and organizations to provide assistance to the FSB in carrying out their assigned duties. Also, individuals and legal entities in Russia that are related to telecommunication services and data communication are obliged to include extra hardware equipment and software or other means for operational and technical access of the data or information to the FSB.[31] Amending the Federal Law on Counterterrorism in 2016 (Federal Law No. 374),[32] the intelligence and secret services have expanded rights in monitoring electronic communication, with legal backing for the interception of personal information. Network operators are also obliged to "keep metadata about all connections, transmissions, and receipts of voice information, written texts, images, sounds, video, and other messages transferred through communications networks" for a period of three years. Moreover, communication companies are required to hand over encryption keys to security agencies on demand for unrestricted access to the plaintext.

India has a complex and distributed regulatory environment with legislations such as the Information Technology Act (IT Act) and Indian Telegraph Act, in addition to the regulations for specific sectors of banking, finance, and telecommunications. Prominently, Section 69 of the IT Act empowers the central and state governments to compel assistance from any "subscriber or intermediary or any person in charge of the computer resource" in decrypting information, and they must extend all facilities and technical assistance to intercept, monitor, or decrypt the information.[33] Moreover, Section 84A of the IT Act grants authority to the Central Government to prescribe the modes or methods of encryption for the secure use of the electronic medium and for the promotion of e-governance and e-commerce.[34] The government invited public comments for its draft national policy on encryption (under Section 84A of IT Act) in 2015. It was withdrawn immediately due to widespread criticism from civil society and privacy activists owing to a few requirements provisioning individual and business users to retain the plaintext information for 90 days, and mandatory registration of encryption products.

Different sectors, particularly, banking, stock markets, and telecommunications, have stipulated requirements for minimum standards of encryption, put up by the respective regulatory body. The Securities and Exchange Board of India prescribes 128 bit encryption standard (using Secured Socket Level Security) for secure transactions between the depository, participants, issuers, and agents.[35] As a minimum security standard, the Reserve Bank of India also mandates Secured Socket Layer for server authentication and the use of client side certificates. It prescribes 128-bit SSL encryption for secure web browsers to server communications, and the encryption of sensitive data, like passwords in transit within the enterprise.[36] The management and administration of encryption products and services is disparate. India needs a comprehensive and coherent national policy for encryption, encompassing all the sectors which deploy encryption products and services. The existing framework is fragmented, and a lot of smartphone applications are exploiting the gaps, particularly Over-The-Top apps and services which deploy stronger encryption for end-to-end communications security. Any further delay in rolling out a national policy for encryption will have severe implications for the economic benefits India wants to derive from the digital economy.

In the case of the European Union (EU), there is no mandate or a prescribed requirement for private entities or individuals to assist governments in decrypting communications or key escrow.[37] In its position paper on encryption, the European Union Agency for Network and Information (ENISA) has resisted the very idea of backdoors in encryption products, or weakening the strength of the encryption algorithms.[38] In the aftermath of the Paris and Brussels terrorist attacks, the debate on expanding the reach of law enforcement to unencrypted data further intensified all across the EU. The views within the EU are also divergent, as a few of the member states are keen to extend the access of law enforcement agencies to personal information in the wake of the increasing use of encrypted messaging apps and other products by the terrorist outfits, and also in the case of lone-wolf attacks.

In France, national intelligence and security services are authorised to intercept and access private communications; but it is subject to specifically defined purposes such as to protect national security, prevent the acts of terrorism or crime in the interest of the economy.[39] In 2016, the French National Assembly rejected a proposed amendment that required mandatory back doors in technology products using encryption,[40] mooted in the wake of the Paris and the Nice terrorist attacks. The government regulates the supply, the import and export of cryptographic means in and from France, either through a declaration or an authorisation process.[41]

The United Kingdom (UK) has had legislation since the early 2000s to enable law enforcement and intelligence officials to lawfully get access to encrypted information, known as "enforced decryption."[42] The Investigatory Powers Law of 2016 brought in a complete overhaul to the legal framework; it allows the government to compel communication providers to remove "electronic protection applied to any communications or data".[43] The UK has also been a victim of a series of terrorist acts, be it in London or Manchester.

Confronting mounting challenges from the intersection of technology and the acts of terrorism or crime, the government is planning to extend the existing Investigatory Powers Law to allow for near real-time surveillance and the removal of encryption.

There are formidable legal and technological challenges in the domestic realm before governments can allow lawful access to encrypted data. Whether it is the idea of backdoors, restricting the strength of the encryption, mandatory technical support in investigations, or key escrows, there is resistance on account of privacy and civil liberties. The rising threats from terrorism and transnational crime are a growing concern for governments as they struggle to strike the right balance between technology innovation and legitimate security concerns. On the international and multilateral front, the challenges are also daunting, particularly in terms of building a global consensus on the rules, regulations, and the specific conditions under which governments can access plaintext, along with appropriate measures to ensure the transparency of such transactions and defining the responsibilities of citizens, government establishments, and industry.

The discourse can move in either of two ways: in the favour of privacy and civil liberties, or in favour of security and intelligence agencies. Nevertheless, this does not stop the security agencies from investing in their cryptanalysis capabilities. Under both the circumstances, whether governments make legislations to allow access to encrypted data or their security agencies covertly hone their cryptanalysis dexterity, the unaddressed concerns over ethical and legal considerations loom large. But certainly, the backdoor approach to circumvent encryption is detrimental to the interests of the consumers and commercial developers which, in turn, jeopardize economic prospects.

In a paper authored by a group of eminent cryptographers, the experts have thoroughly evaluated the proposed idea of exceptional access to communications in the context of the complex and globalized information infrastructure. The experts have found this idea to impose grave security risks, imperil innovation, and raise issues for human rights and international relations.[44] Implementation itself is practically challenging because, in such an arrangement, security credentials would have to be retained by the platform provider, law enforcement agencies, or some other trusted third party. An intrusion into any of these could compromise the security of the entire information infrastructure. The risks from malicious insiders also run high.

Apprehension has also arisen in the wake of attacks on the US Government Office of Personnel Management (OPM), where numerous federal agencies lost their sensitive data owing to the insecure infrastructure of the OPM. A compromise at a trusted institution can inflict serious damage.

Also, nation states have multiple law enforcement agencies, and their jurisdictions are spread across the state, ministries, departments and other administrative divisions. Exceptional access can complicate the impending issue of legal jurisdiction.[45] This also throws open inevitable challenges for the vendors and technology developers to comply with difference legal and licensing regimes across different nation states. The difficulty in reaching a global consensus, as witnessed from experience of the UNGGE, either for the harmonisation of laws or interoperable regimes, further complicates the impending issues. Encryption is practically widespread, be it for protecting stored files and full disk encryption or to secure web browsing, messaging, or transactions. The discourse must move forward from the baseline understanding that there is no substitute for strong encryption, and the way encryption systems have developed as open-source technology, they are in the reach of almost everyone.

In democratic countries, technology users have become quite aware and vigilant of their privacy online. For obvious reasons, authoritarian regimes have stringent controls on data without paying much heed to the privacy of users. As a serious detriment to surveillance and investigative efforts, governments, even in democratic setups have chosen to adopt laws which could give them unrestricted access to the content, either unencrypted or decrypted. As a zero sum game, the tilt, as of now, is certainly towards security, which clearly infringes upon the privacy of individuals. Governments have to find the right balance between privacy and public safety in close consultation with the public, civil society, and industry, notwithstanding national security concerns which have international diplomatic and technical dimensions.

Encryption Export Controls under the Wassenaar Arrangement

The Wassenaar Arrangement, the voluntary agreement comprising forty two nations, was set up in 1996 to control the sale and export of conventional arms and goods, or technologies having dual-use. Formally known as the "Wassenaar Arrangement on Export Controls for Conventional Arms and Dual-Use Goods and Technologies", it aims to promote responsibility and transparency in the global arms trade. In effect, it is an arms control regime in the post-Cold War era to control the exports of conventional munitions and dual-use goods and technologies varying from "Bombs, torpedoes, rockets, missiles, and other explosive devices;" to "[c]hemical or biological toxic agents;" and "[n]uclear [p]ower generating equipment."[46] Over the years, the controls list has eventually grown to nine categories including technologies related to information and cybersecurity. The implementation of controls is carried out through national legislation once the participating states agree to maintain the controls. As a compliance requisite, the vendors require a license to export a product using cryptography beyond certain strength.

The export of cryptography has been controlled ever since the end of World War II, primarily for national security reasons, because the applications of cryptography then were largely limited to the military domain. Export controls were applicable to cryptography beyond certain strength, defined by the algorithm and length of the encryption key. Diffie-Hellman key exchange and RSA algorithms brought a paradigm shift; the use of encryption expanded to commercial and consumer realms. The demand for cryptographic applications in the commercial sphere grew for securing financial transactions. In 1996, expanding commercial applications forced the USA to transfer the administration of encryption export controls from the Munitions List category to the Commerce Control List.

Classifying cryptographic systems as either military or civilian based on the strength of the key or encryption algorithm was not feasible as the distinction between both is not as simple as in the case of fire arms or artillery weapons. The same applied to differentiating the end users as either military or civilian.

Recognising the strategic imperatives of strong cryptographic functions, the ultimate goal of such export control regimes is to keep it out of the reach of nation states that are a probable target of signals intelligence,[47] or pose a threat to international peace and stability. Earlier, when cryptography was implemented in hardware, it was comparatively easy to enforce export controls and ensure compliance. Once the cryptographic functionality was shifted from hardware to software products, it became difficult to enforce the existing controls and compliance mechanisms since software products could be distributed or replicated anywhere all across the globe. Export controls also had detrimental effects on the commercial sector, both through loss of business opportunity as well as access to technology for countries outside the Wassenaar Arrangement.

The effectiveness and utility of this export control regime is generally questioned on the pretext of the proliferating use of strong encryption in most countries, even in those outside the Wassenaar Arrangement or in the backdrop of the continued legislative efforts for backdoors or restrictions on algorithms and key length. In addition, the Open-source-software movement has also made it easy to develop software products implementing encryption,[48] with instant access to source codes and other development toolkits. India was out of Wassenaar Arrangement until 2017, when it became its 42nd member,[49] putting an end to the technology denial regime. This will give India seamless access to the cutting-edge of technology in cryptography. Diplomatic strength and India's strong non-proliferation credentials have helped India surpass this challenge. However, an imminent threat arises from Quantum computing and information sciences, which are slated to alter the mathematical and theoretical assumptions on which classical cryptography rests. Encryption key distribution over a quantum channel at the one end can help build hack-proof networks; but quantum computing can render some of the algorithms weak and insecure.

The Challenge of Quantum Computing for Encryption

The primary purpose of an encryption algorithm is to protect sensitive and confidential data or information, both in transit or stored on a memory device or computer, or in any other electronic or printed format. Encryption algorithms are also standardised or recommended by governments for wider usage in securing their secrets, be it Data Encryption Standard (DES), Triple DES, or Advanced Encryption Standard (AES). They underpin the security of IT and information systems and digital data communication, although they have a limited lifespan over which they are deemed to be safe to use against prominent attacks. As encryption algorithms improve, the mathematical and computational ways and means of cryptanalysis also advance, rendering encryption algorithms unsafe over time.

Scientific advancements in quantum mechanics and experiments with "superposition" phenomena have unleashed their vast potential for computing, paving the way for quantum computers. These computers can store information, not just in the two states of 0 or 1 as classical computers do, but also as superposition of these values, in qubits. These properties help quantum computers in executing complex computations which were beyond the capacity of classical computers, for a wide range of applications such as drug discovery, molecular modelling, weather forecasting, artificial intelligence, and most important, cryptography.

Public-key cryptographic systems and key exchange protocols such as RSA, Elliptic curve, or Diffie-Hellman key exchange,[50] are vital to the security of present day digital communication for the exchange of secret keys, particularly in the Internet era. Also, as discussed in the previous section, public-key cryptographic systems were a prime enabler of the burgeoning commerce over the Internet and other security products for commercial and individual purposes. Public-key cryptographic systems are based on mathematical assumptions. For instance, the RSA, assumes that it is mathematically easy to multiply two large prime numbers, but computationally intensive to factorize the product of prime numbers back into two prime numbers. Quantum computers, overcoming the limitations of classical computers, are likely to solve mathematical enigmas and break these encryption systems. They could also solve the integer factorization problem using the two-and-a-half decade old Shor's algorithm,[51] and even break the Diffie-Hellman or Elliptic Curve Diffie-Hellman key exchange protocols, designed on the principles of integer factorization or the discrete log problem.[52] Intelligence agencies and militaries have keen interests in breaking cryptographic systems, which also explains their deep interest and aggressive funding for both theoretical and experimental research programs in quantum computing. The Pentagon and the US intelligence community have been supporting such research efforts.[53] Quantum Computers will certainly be used to break encrypted data sets as well as to decrypt information intercepted over the years, but was inaccessible until now.

Quantum computing will possibly render some of the prevalent secret-key or public-key cryptographic systems weak. Some of them would need to be fortified with longer encryption keys which may be impractical in implementation.[54] The focus of further research has now shifted towards redesigning or identifying algorithms that are resistant to attacks by both classical and quantum computers and these efforts of the international community are now known as "Post-Quantum Cryptography". The quest for new algorithms, resistant to attacks using quantum computers has already begun, and a few of the proposals—such as Lattice-based cryptography,

Multivariate-quadratic-equations cryptography, Hash-based cryptography, and Code-based cryptography—are under consideration.[55]

The NIST had also invited proposals for quantum-resistant cryptographic algorithms for new public-key crypto standards.[56] The goal is to develop cryptographic systems that are secure against both quantum and classical computers, and to improve their efficiency and usability.[57] The European Telecommunications Standards Institute (ETSI) also recognises the risks to the security and privacy from business, ethical, and legal perspectives.[58] Quantum-safe cryptography is essential to protect government and military communications, secure financial and banking transactions, medical data and healthcare records, and also to safeguard privacy and personal data. ETSI has also rolled out initiatives to assess and make recommendations for quantum-safe cryptographic primitives and protocols, in the form of the Industry Specification Group,[59] and a series of workshops.[60] Global efforts have gained pace to figure out the measures to secure information encrypted with prevalent cryptographic systems and encryption algorithms, mitigating the risks from quantum computing.

There is a continuous race between the disciplines of cryptography and cryptanalysis, both with their respective advantages and risks. Information security, as a broader practice, relies heavily on cryptographic functions, encryption being one of the prominent ones. Historically, cryptography has been a subject of military and strategic importance; but ever since World War II, the technology, export controls, and policy decisions surrounding encryption have influenced both international relations and the security calculus.

Strong encryption is vital to the security of information systems which underpin our modern societies, economies, and nation states. Protecting cryptographic systems and encryption algorithms from any political interference is of utmost importance, particularly in the backdrop of their widening use in a number of mobile platforms. Moreover, the advancements in other spheres of computer and information sciences, particularly those in artificial intelligence and big data analytics, are also changing the way cyber and information systems are protected against known as well as unknown threats.

Cybersecurity in the Age of Artificial Intelligence

As a burgeoning discipline in the realm of computer science, Artificial Intelligence (AI) enables machines to think, learn, and respond in ways quite similar to the human abilities of speech, facial, object, or gesture recognition, problem solving, reasoning, and perception. AI unfolds a plethora of real-world applications, which are as sparse as autonomous vehicle, social humanoids, intelligent personal assistants, and even autonomous weapons which are capable of executing military missions on their own. It has also made inroads in industrial automation and decision support systems, heavily complementing or augmenting human abilities. AI empowers computer systems to learn on their own, rather than depending upon any pre-programmed set of instructions or pre-defined behavioural algorithms. They can learn from their interactions or experiences, and enhance their capabilities, knowledge and skills.[61] AI, in essence, considerably enhances the ability of computer systems to learn from their experiences over time, using 'machine learning' techniques. Such machines are quite capable of reasoning, perceiving relationships and analogies, solving problems, and are even adept at using natural languages for interaction.

Technology firms all across the globe are backing research in AI, be it Apple, Amazon, Google, Facebook, IBM, or Microsoft in the USA, or Baidu, Alibaba or Tancent in China. AI also is of profound interest to governments and armed forces. AI is unveiling path-breaking applications in fields as diverse as healthcare and life sciences to data analysis, cybersecurity and finance. Private enterprises are experimenting with AI for IT functions. However, they are quite certain to extend it to marketing, customer service, finance, human resources, strategic planning, and other corporate functions.[62] AI is also helping to strengthen the technology and practices of cybersecurity for enhanced protection against sophisticated threat actors. AI automates various processes, such as malware and anomaly detection, risk analysis and response.

AI as a standalone technology may not be the right solution, but merged with human intelligence and expertise, it can augment existing capabilities for better defences and responses. Magnifier, a product of Palo Alto Networks based on behavioural analytics to model network behaviour, improves the detection of stealth threats and the identification of targeted attacks, malicious insiders, and malware.[63] Alphabet, the parent company of Google, has also rolled out a dedicated business unit for cybersecurity to couple AI with swathes of data and computing power for the quick and precise detection of threats. A lot of technology companies are now innovating in this segment of cybersecurity practice, be it RazorSecure targeting the aviation, rail and automotive markets,[64] Cylance applying artificial intelligence, algorithmic science, and machine learning for advanced threat prevention,[65] JASK using big data analytics alongside AI,[66] or Sovereign Intelligence using AI to gather and analyse data from non-traditional data sources, like Dark web for an insightful view on probable external threats.[67]

With the influx of corporate interests in harnessing the benefits of AI for cybersecurity, novel applications can accelerate or automate incident detection, response, and remediation while helping professionals in correlating high volumes of security alerts. It can also assist in detecting software vulnerabilities or configuration errors in the source code. There is no dearth of possibilities of applications, which could be simple like spam filtering, botnet detection, and user authentication, or complex like fraud detection, threat intelligence, and incident forecasting. AI aided solutions for cybersecurity will certainly cut down the number of incidents as well as detection time, probably from hundreds of days to a few days or even to a few hours.[68] However, investments in AI and machine learning have to be sustained for the long term.

Data Security and Privacy Concerns

The data-driven digital age has amplified the ambivalent tension between the public's desire for privacy and the need for security. States authorise monitoring and surveillance to combat the perils of terrorism, extremism, radicalisation, espionage, and a host of other threats from both state and non-state actors. In democratic political systems, civil liberties are generally enshrined in the Constitution or the social fabric. In the case of India,[69] privacy is also increasingly being recognised as a fundamental right. Amidst these changes, the state has to optimise adequate security and protection with wider economic, business and social interests. National security and law enforcement requirements compete with the requirements of the free flow of information, free speech, online anonymity, and private communications.

In the wake of state-led surveillance and instances of personal data harvesting from online activities of users, the voices for privacy have gained prominence towards strong data protection frameworks. User behaviour in online platforms, particularly in the voluminous social media and e-commerce segment, is susceptible to be used for targeted marketing or targeted propaganda as part of election campaigns, or for that matter, foreign intervention in influencing electoral outcomes. The issue, *per se*, came to the limelight in the aftermath of the alleged Russian involvement in the US Presidential elections in 2016, and also with the unravelling of Facebook-Cambridge Analytica scandal. These instances have also made the users inquisitive, and aware of their online activities, rights and further usage of their personal data.

Data protection, essentially, targets safeguarding citizens from the perils of misuse of personal data owing to loss, alteration, theft, unauthorised access to, or unintended use, by affixing responsibilities of the government, companies as well as individuals handling such data. The considerations of security, foreign surveillance, law enforcement investigations are the prime drivers for governments to enact legislations entailing data localisation within their political boundaries.

In May 2018, the European Union enforced the General Data Protection Regulation (GDPR), strengthening the data rights of the residents of the European Union and harmonising data protection laws all across the member states. The GDPR intends to bring transparency in the way personal information is collected and used, giving more control and rights to the individual over their personal data, under a unified regulatory environment.[70] In accordance with the global changes, India is also building its data protection framework to enable a strong data privacy regime. The draft Personal Data Protection Bill 2018, submitted to the government for perusal, is an outcome of the deliberations and consultations led by a committee of experts under the chairmanship of Justice B.N. Srikrishna. Walking a tightrope between India's developmental needs and privacy requisites, the bill aspires to build a legislation which will protect the privacy of the individuals, ensure their autonomy, and simultaneously allow data flows to create a free and fair digital economy.

Terming it as "A Fourth Way to Privacy, Autonomy and Empowerment", the committee of experts has taken an approach which is distinct from the USA, the EU and China; it has tried to present a fourth path.[71] With disparate sector-specific regulations already in place—such as in the form of Information Technology Act, The Indian Copyright Act or Credit Information Companies Regulation Act—the foremost step for India would be to overcome the multi-agency regulatory structure to a simpler national authority which can ensure compliance and effective implementation. Given the sensitive nature of the data which exchanges hands among different vendors and suppliers, the respective regulators from healthcare and telecom sectors,[72] have also floated consultation papers to garner views and representation for regulations pertaining to privacy, security, and ownership of data in the specific domains.

The state reserves the right to protect its interests and safeguard its sovereignty; but excessive emphasis on protection and stringent regulations can stifle innovation in data sciences and technologies. For instance, data localisation laws may not be fit for price sensitive markets where technology providers build high availability services with lower investments in capital and equipment. The main driver for lower costs is the integration of information systems with global spread of data storage and processing. Governments have to balance their legitimate and reasonable security concerns with the requirements of economic and societal development. The approaches and laws for data protection have subjective applicability and relevance, as the requirements, digitisation, and technology maturity, etc. vary across every nation state. Taking a constrained view on security, sovereignty and protection could also be counter-productive.

India's Technology Challenges

India had been out of Wassenaar Arrangement until 2017, when it became its 42nd member. India hitherto not had access to military-grade and sensitive technologies governed by the Wassenaar Arrangement. It is of utmost importance now for India to leverage the access to Wassenaar Arrangement controlled technology, with cryptographic products, information security software and technology being one of the prominent ones. Poised at the brink of becoming a digital and knowledge based economy, India needs business and investments in the technology sector. Encryption ensures the security of sensitive data and information, and strong encryption is important to address data security and privacy concerns. Weaker policy, regulatory measures, or inferior technology can seriously undermine the growth prospects of a digital economy.

A number of India's security and law enforcement concerns stem from the expanding use of encryption for data storage and messaging applications. Security features like end-to-end encryption in messaging services enhance privacy for the users and but also facilitates their misuse by terrorists and criminals or anti-national elements. Investigations in the cases of crime and terrorism turn out to be extremely challenging when the seized digital devices have hard drive encryption. India's susceptibility to the acts of terror, left wing extremism, communal violence and fake information triggers security concerns. Government may want access to content in order to facilitate investigations and prosecution; but technology developers are not willing to undermine the privacy of their users.

India is strategically important for technology companies, especially to those in the practice of information and cybersecurity. Private enterprises are also increasing their Research and Development resource base in India. India is also witnessing an exponential rise in technology start-ups which are going to lead innovation in the technology segment of cybersecurity. The opportunities and benefits are plenty, but are not without risks and challenges.

India fares average in the surging competition for encryption, AI, and quantum technology development. The research output from India in terms of research papers in reputed international journals ranks at 7 for AI. Most importantly, India lacks a clearly stated policy document or vision statement for the development of these technologies. There are few efforts, but they appear to lack coherence. On the academic front, the Indian Institute for Science hosted the Centre for Quantum Information and Quantum Computation with funding from the Department of Science and Technology (DST) from 2010 to 2015. The Quantum Information and Computation (QIC) Group at the Harish-Chandra Research Institute is also involved in cutting-edge research on quantum algorithms, quantum communication, quantum cryptography, and the theory of entanglement.[73] The Tata institute for Fundamental Research has also built the Quantum Measurement and Control Laboratory.[74] The DST has also initiated and invited proposals under a directed research programme on "Quantum Information Science and Technology (QuST)".[75] The Department of Defence Production has constituted a 17-member task-force to study the use of AI for both military applications and technology-driven economic growth in February 2018.[76]

AI, for certain, has caught the eye of the government; but India has to grow its own technical competence in these fast-paced disciplines as foreign dependence would be unproductive. Delays in putting the right policy frameworks in place run the acute risk of pushing India to the early majority, late majority, or even towards laggards in the technology adoption bell curve, severely limiting its ability to draw the economic advantage.

Maturing a step further from theoretical research, India needs more experimental facilities to build the prototypes. Sustained funding will ensure continuity in research programs. The Centre for Quantum Information and Quantum Computation, for instance, ceased to function once the funding from DST was over in 2015. The private sector is also expected to invest and inculcate innovation. However, American entities dominate the existing landscape in India, with the likes of Accenture and Microsoft establishing centres for innovation. Enterprises of Indian origin are yet to make inroads, without which the requisite ecosystem will be incapacitated to deliver the desired results. Moreover, these technology segments are truly multi-disciplinary; so integrating expertise from computer science, physics, electronics, materials, data science, information science, and mathematics is vital to success. India has to contemplate global developments in perspective in order to develop a research oriented ecosystem harnessing technologies for cybersecurity. However, it should make sure that it does not end up remaining a mere consumer in this expansive market.

NOTES

1. These backdoors could be exploited to gain access to the networks of the targeted organization, such as research laboratories, defence contractors, embassies, government offices, corporations, and so on, where the software or hardware has been deployed, posing a serious threat to information pertaining to the government.
2. Bruce Schneier, "History of the First Crypto War", Schneier on Security, 22 June 2015, at https://www.schneier.com/blog/archives/2015/06/history_of_the_.html, accessed on 20 January 2018.
3. Ibid.
4. Andrea Peterson, "The 'Crypto Wars' of the 1990s are brewing again in Washington", *The Washington Post*, 10 September 2015, at https://www.washingtonpost.com/news/the-switch/wp/2015/09/10/the-crypto-wars-of-the-1990s-are-brewing-again-in-washington/?utm_term=.84f318fa99ab, accessed on 20 January 2018.
5. Cryptanalysis deals with deciphering encrypted communications without knowing the encryption keys. There are many cryptanalytic techniques, such as Ciphertext-only attack, Known-plaintext attack, Chosen-plaintext attack, etc.
6. To protect the data for a cash-dispensing system that IBM developed for Lloyds Bank in the United Kingdom.
7. "Cryptography for a Connected World", IBM at 100, at http://www-03.ibm.com/ibm/history/ibm100/us/en/icons/cryptography/, accessed on 20 January 2018.
8. Although the algorithm accepted 64 bit key as input, the remaining 8 bits were used for parity checking, which had no effect on the security of DES. For details see, Charlie Kaufman, Radia Perlman, Mike Speciner, Network Security (Pearson Education, 2005), p. 63; and Arthur Sorkin, "Lucifer, A Cryptographic Algorithm", Lawrence Livermore Laboratory, April 1983, at http://fuseki.com/lucifer.pdf.

9. Joe Gargiulo, "S-Box Modifications and their Effects in DES-like Encryption Systems", SANS Institute InfoSec Reading Room, July 2002, at https://www.sans.org/reading-room/whitepapers/vpns/s-box-modifications-effect-des-like-encryption-systems-768; and "Data Encryption Standard", at http://www-math.ucdenver.edu/~wcherowi/courses/m5410/des.pdf.
10. Darren Pauli, "Leaked NSA Docs Suggest Dual_EC_DEBG Backdoor", *IT News*, 11 September 2013, at http://www.itnews.com.au/News/356751,leaked-nsa-docs-suggest-dualecdrbg-backdoor.aspx, accessed on 20 January 2018.
11. National Institute of Standards and Technology, "NIST Removes Cryptography Algorithm from Random Number Generator Recommendations", April 21, 2014, at https://www.nist.gov/news-events/news/2014/04/nist-removes-cryptography-algorithm-random-number-generator-recommendations, accessed on 20 January 2018.
12. James Ball, "How US and UK spy agencies defeat internet privacy and security", *The Guardian*, 06 September 2013, http://www.theguardian.com/world/2013/sep/05/nsa-gchq-encryption-codes-security, accessed on 20 January 2018; and Adam Clark, "The NSA can beat almost any type of encryption", Gizmodo, 5 September 2013, at http://gizmodo.com/the-nsa-can-crack-almost-any-type-of-encryption-1258954266, accessed on 20 January 2018.
13. Bruce Schneier, "How the NSA targets users' online anonymity", *The Guardian*, 04 October 2013, at http://www.theguardian.com/world/2013/oct/04/tor-attacks-nsa-users-online-anonymity/, accessed on 20 January 2018.
14. Jacob Appelbaum, Judith Horchert, and Christian Stöcker, "Catalog Reveals NSA Has Back Doors for Numerous Devices", *Spiegel Online*, 29 December 2013, at http://www.spiegel.de/international/world/catalog-reveals-nsa-has-back-doors-for-numerous-devices-a-940994.html, accessed on 20 January 2018; and Kim Zetter, "NSA Hackers get the 'Ungettable' with Rich Catalog of Custom Tools", *Wired*, 30 December 2013, at http://www.wired.com/threatlevel/2013/12/nsa-hacking-catalogue/, accessed on 20 January 2018.
15. Eric Geller, "A complete guide to the new 'Crypto Wars'", *The Daily Dot*, 05 May 2016, at https://www.dailydot.com/layer8/encryption-crypto-wars-backdoors-timeline-security-privacy/, accessed on 24 January 2018.
16. James Ball, Julian Borger and Glenn Greenwald, "How US and UK spy agencies defeat internet privacy and security", *The Guardian*, 06 September 2013, at https://www.theguardian.com/world/2013/sep/05/nsa-gchq-encryption-codes-security, accessed on 24 January 2018.
17. Jon M. Peha, "Encryption Policy Issues", Carnegie Mellon University, October 1998, at http://repository.cmu.edu/cgi/viewcontent.cgi?article=1037&context=epp, accessed on 24 January 2018.
18. Cara McGoogan, James Titcomb and Charlotte Krol, "What is WannaCry and how does ransomware work?" *The Telegraph*, 18 May 2017, at http://www.telegraph.co.uk/technology/0/ransomware-does-work/, accessed on 24 January 2018.
19. James A. Lewis, Denise E. Zheng and William A. Carter, "The Effect of Encryption on Lawful Access to Communications and Data", Centre for Strategic and International Studies, February 2017, p. IV.
20. "Perfect Forward Secrecy", Telegram, at https://core.telegram.org/api/pfs, accessed on 21 January 2018.
21. "Encryption and Export Administration Regulations", Bureau of Industry and Security of the U.S. Department of Commerce, 15 August 2017, at https://www.bis.doc.gov/index.php/policy-guidance/encryption, accessed on 21 January 2018.

22. "Decrypting the Encryption Debate: A Framework for Decision Makers", Committee on law Enforcement and Intelligence Access to Plaintext Information, Computer Science and Telecommunications Board, Division of Engineering and Physical Sciences, 2018, The National Academies Press, Washington D.C., at https://www.nap.edu/read/25010/chapter/1, accessed on 21 January 2018.
23. Ibid., pp. 8–9
24. Bill to require the provision of data in an intelligible format to a government pursuant to a court order, and for other purposes, The Senate of the United States, Discussion Draft at the 114th Congress, BAG16460, at https://www.burr.senate.gov/imo/media/doc/BAG16460.pdf, accessed on 21 January 2018.
25. "China Releases Draft Encryption Law for Public Comment", Covington, 09 May 2017, at https://www.cov.com/-/media/files/corporate/publications/2017/05/china_releases_ draft_encryption_law_for_public_comment.pdf, p. 1.
26. Article 20 of China's Cryptography Law, Translation by the European Chamber, at https://www.steptoe.com/assets/attachments/4966.pdf, p. 5.
27. Article 16 of China's Cryptography Law, Translation by the European Chamber, at https://www.steptoe.com/assets/attachments/4966.pdf, p. 4.
28. "China Revises Rules on Commercial Encryption Products", Covington, 15 October 2017, at https://www.cov.com/-/media/files/corporate/publications/2017/10/china_revises_rules_on _commercial_encryption_products.pdf, p. 1.
29. "China: Provisions on Decryption of Communications in Anti-Terrorism Law", Global Legal Monitor, Library of U.S. Congress, 17 February 2016, at http://www.loc.gov/law/foreign-news/article/china-provisions-on-decryption-of-communications-in-anti-terrorism-law/, accessed on 28 January 2018.
30. "Russian Laws and Regulations: Implications for Kaspersky Labs", at https://www.wired.com/images_blogs/dangerroom/2012/07/Russian-Laws-and-Regulations-and-Implications-for-Kaspersky-Labs.pdf
31. Ibid.
32. Also known as the Federal Law on Counterterrorism Select Legislative Acts of the Russian Federation Concerning the Creation of Additional Measures Aimed at Countering Terrorism and Protecting Public Safety.
33. "The Information Technology (Amendment) Act, 2008", Ministry of Law and Justice, Government of India, 05 February 2009, at http://www.meity.gov.in/writereaddata/files/it_amendment_act2008%20%281%29_0.pdf, p. 12.
34. Ibid., p. 17.
35. "Trading Software and Technology", Securities and Exchange Board of India, at https://www.sebi.gov.in/sebi_data/commondocs/chapter2trading_p.pdf, pp. 3–6.
36. "Internet Banking in India: Guidelines", Reserve Bank of India, 14 June 2001, at https://www.rbi.org.in/scripts/NotificationUser.aspx?Id=414&Mode=0, accessed on 28 January 2018.
37. Government Access to Encrypted Communications, The Law Library of Congress, Global Legal Research Centre, May 2016, at https://www.loc.gov/law/help/encrypted-communications/gov-access.pdf.
38. "ENISA's Opinion Paper on Encryption: Strong Encryption Safeguards our Digital Identity", The European Union Agency for Network and Information Security, December 2016, at https://www.enisa.europa.eu/publications/enisa-position-papers-and-opinions/enisas-opinion-paper-on-encryption, p. 5.

39. Government of France, Interior Security Code, art. L811-3, archived at https://perma.cc/Z32U-CVJA.
40. Jeff John Roberts, "France Rejects Backdoors Law to Defeat Encryption", *Fortune*, 13 January 2016, at http://fortune.com/2016/01/13/france-encryption/, accessed on 30 January 2018.
41. "Encryption Control", ANSSI (The National Cyber security Agency of France), at https://www.ssi.gouv.fr/en/regulation/cryptology/, accessed on 02 February 2018.
42. Regulation of Investigatory Powers Act 2000, at https://www.legislation.gov.uk/ukpga/2000/23/contents, accessed on 02 February 2018 and David Anderson, Report of the Investigatory Powers Review – Presented to the Prime Minister pursuant to section 7 of the Data Retention and Investigatory Powers Act 2014, June 2015, at https://perma.cc/N4UN-UE7F, accessed on 02 February 2018.
43. "UK Surveillance Powers Explained", *BBC*, 05 November 2015, at http://www.bbc.com/news/uk-34713435 accessed on 02 February 2018; and Kieren McCarthy, "UK's new Snoopers' Charter just passed an encryption backdoor law by the backdoor", *The Register*, 30 November 2016, at https://www.theregister.co.uk/2016/11/30/investigatory_ powers_ act_ backdoors/, accessed on 02 February 2018.
44. Harold Abelson et al., "Keys under doormats: mandating insecurity by requiring government access to all data and communications", *Journal of Cybersecurity*, Volume 1, Issue 1, September 2015, pp. 69–79, at https://doi.org/10.1093/cybsec/tyv009.
45. Ibid.
46. "Wassenaar: Turning Arms Control into Software Control", 25 May 2015, see http://www.internetgovernance.org/2015/05/25/wassenaar-turning-arms-control-into-software-control/, accessed on 02 February 2018.
47. n. 22, Chapter 9.
48. "Roll Your Own Crypto Services (Using Open Source and Free Cryptography)", SANS Institute InfoSec Reading Room, 24 January 2002, at https://www.sans.org/reading-room/whitepapers/vpns/roll-crypto-services-using-open-source-free-cryptography-758, accessed on 04 February 2018.
49. Wassnaar Arrangement, "India becomes 42nd WA Participating State", 08 December 2017, http://www.wassenaar.org/india-becomes-42nd-wa-participating-state-8-dec-2017/, accessed on 04 February 2018.
50. National Institute of Standards and Technology, Report on Post-Quantum Cryptography (NIST Interagency Report 8105, April 2016), p. 1, at http://csrc.nist.gov/publications/drafts/nistir-8105/nistir_8105_draft.pdf.
51. "Shor's Algorithm", Lecture by Ryan O'Donnell, Scribe by Sidhanth Mohanty, Carnegie Mellon University, 07 October 2015, at https://www.cs.cmu.edu/~odonnell/quantum15/lecture09.pdf.
52. n. 50.
53. Sharon Weinberger, "Why Google and the Pentagon want Quantum Computers", *BBC*, November 18, 2014, http://www.bbc.com/future/story/20130516-big-bets-on-quantum-computers, accessed on 05 February 2018.
54. "New encryption methods strengthen current cyber security efforts and establish a platform for the secure communication of quantum computers", Stevens Institute of Technology, at http://research.stevens.edu/post-quantum-cybersecurity, accessed on 05 February 2018.
55. Daniel J. Bernstein, "Introduction to post-quantum cryptography", in Daniel J. Bernstein, Johannes Buchmann and Erik Dahmen (eds.), *Post-Quantum Cryptography* (Berlin: Springer, 2009), p. 1.

56. n. 50.
57. n. 55, p. 11.
58. "Quantum-Safe Cryptography", European Telecommunications Standards Institute, at http://www.etsi.org/technologies-clusters/technologies/quantum-safe-cryptography, accessed on 05 February 2018.
59. "Quantum Computing and the risk to security and privacy", European Telecommunications Standards Institute, at http://www.etsi.org/images/files/ETSITechnologyLeaflets/QuantumSafeCryptography.pdf, accessed on 05 February 2018.
60. "4th ETSI/IQC Workshop on Quantum-Safe Cryptography", European Telecommunications Standards Institute, at http://www.etsi.org/index.php/news-events/events/1072-ws-on-quantumsafe-2016, accessed on 05 February 2018.
61. R. L. Adams, "10 Powerful Examples of Artificial Intelligence in Use Today", *Forbes*, 10 January 2017, https://www.forbes.com/sites/robertadams/2017/01/10/10-powerful-examples-of-artificial-intelligence-in-use-today/#1e4867ad420d, accessed on 17 February 2018.
62. Based on "Global Trend Study on AI" conducted by Tata Consultancy Services, https://www.tcs.com/artificial-intelligence-to-have-dramatic-impact-on-business-by-2020, accessed on 19 February 2018.
63. "Magnifier Behavioural Analytics", 20 February 2018, Palo Alto Networks, at https://www.paloaltonetworks.com/resources/datasheets/magnifier, accessed on 22 February 2018.
64. "Protection on the move: Cyber Security for Aviation, Rail and Automotive", Razor Secure, at https://www.razorsecure.com/, accessed on 22 February 2018.
65. "Cylance Protect: Artificial Intelligence Endpoint Security", Cylance, at https://www.cylance.com/en_us/products/our-products/protect.html, accessed on 22 February 2018.
66. "Modernizing Security Operations", Jask, at https://jask.ai/why-jask/, accessed on 22 February 2018.
67. "AI Driven Intelligence for Enterprise Risk", Sovereign Intelligence, at https://www.sovereign.ai/, accessed on 22 February 2018.
68. Nicholas Fearn, "How AI will underpin cyber security in the next few years", *Computer Weekly*, February 2018, at http://www.computerweekly.com/feature/How-AI-will-underpin-cyber-security-in-the-next-few-years, accessed on 22 February 2018.
69. The Supreme Court of India has stated that the "right to privacy is protected as an intrinsic part of the right to life and personal liberty under Article 21 of the Constitution and as a part of the freedoms guaranteed by Part III of the Constitution."
70. "What does the General Data Protection Regulation (GDPR) govern?" European Commission, at https://ec.europa.eu/info/law/law-topic/data-protection/reform/what-does-general-data-protection-regulation-gdpr-govern_en, accessed on 25 July 2018.
71. "A free and fair digital economy: Protecting Privacy, Empowering Indians", Committee of experts under the chairmanship of Justice B.N. Srikrishna, Report submitted to the Ministry of Electronics and Information Technology, Government of India, at http://meity.gov.in/writereaddata/files/Data_Protection_Committee_Report.pdf, pp. 13-14.
72. "Consultation Paper on Privacy, Security and Ownership of the Data in the Telecom Sector", Telecom Regulatory Authority of India, 09 August 2017, at https://www.trai.gov.in/sites/default/files/Consultation_Paper%20_on_Privacy_Security_ownership_of_data_09082017.pdf.
73. Quantum Information and Computation Group at the Harish-Chandra Research Institute, at http://www.hri.res.in/~qic/, accessed on 24 February 2018.

74. Quantum Measurement and Control Laboratory at Tata Institute of Fundamental Research, at http://www.tifr.res.in/~quantro/, accessed on 24 February 2018.
75. "Detailed Call For Proposals (CFP) under ICPS programme", Ministry of Science and Technology, Government of India, at http://dst.gov.in/sites/default/files/QuST%20-%20CFP1.pdf
76. Pranav Mukul, "Task force set up to study AI application in military", *The Indian Express*, 03 February 2018, at http://indianexpress.com/article/technology/tech-news-technology/task-force-set-up-to-study-ai-application-in-military-5049568/, accessed on 19 February 2018.

CHAPTER 7

Public-Private Partnership in Cybersecurity: Opportunities and Challenges

The National Cyber Security Policy of the Government of India, released in July 2013, has been the primary guiding document for governmental efforts under the broader purview of a secure and resilient cyberspace for citizens, businesses, and the government. The policy underscores the imperatives of collaborative engagement in both technical and operational domains, to enhance the security of cyberspace. The government is cognizant of the role of the private sector in this domain, as private entities share the responsibility of deploying and maintaining vast portions of the information systems, computer networks, and information infrastructure across the country. With deregulation, more focus on privatisation, and the phenomenon of globalization over the last three decades, private enterprises and entities were tightly knit into the economic and security architecture of India. In some critical sectors, like communications, banking, and aviation, private sector leads the industry segments, both in terms of investments and innovation. In cybersecurity, the nature of threats, resource requirements, investments, products development, and a multi-disciplinary character become unattainable or unmanageable without coordination and collaboration among the governments and the private sector.

The National Cyber Security Policy, therefore, calls for an effective partnership and collaborative engagements between the public and private entities to attain the objectives laid out in the policy document. Public-Private Partnership (PPP) is critical in tackling cyber threats, and it is also a key component of the National Cyber Security Policy envisioned to enhance India's cybersecurity landscape. PPP based models and solutions could be well-suited and effective in niche areas which require diverse expertise and scarce skill-sets to achieve a common goal. Cybersecurity and critical infrastructure protection are one such area where PPP can propel innovation, leading to value creation for the private sector and better security products or services for the government—in essence drawing mutual benefit for both.

The term "Public-Private Partnership" describes a "spectrum of possible relationships between the public and private sectors for the cooperative provision of infrastructure development and the associated services."[1] In terms of value proposition, private sector participation could bring the essential technical and managerial expertise, help improve operating efficiency as well as attract capital investment and better consumer care or satisfaction. In India, PPP based projects have been conceived and executed successfully in the energy and infrastructure sectors, such as highways, airports, electricity generation, and distribution, to name a few.

Cybersecurity assumes topmost priority, particularly in the wake of the government's colossal initiatives for e-governance, digital inclusion, and identification, among others, envisioned to transform India into a digitally empowered society and a knowledge economy.[2] As the reach and extent of services, personal information, and sensitive data expands in the digital domain, the general management and regulation of cybersecurity measures would require a close partnership between the government as a guardian of national security, and the private sector as a provider of infrastructure and technology. India has functioning PPP models in different walks of governance, be it civil aviation, energy and utilities, or roads and infrastructure development. The telecom sector in India is primarily led by private players. Banking and financial services, the core of the economic system, has a vibrant engagement of the private sector, be it in providing core banking and financial services or running the stock exchanges. Cybersecurity, as a non-traditional security domain, would require a non-traditional approach to problem solving, and PPP in this case, could help provision solutions to many open problems.

Working together on these critical problems, especially when the private sector is a significant part of the Critical Infrastructure (CI) and Critical Information Infrastructure (CII), is all the more important. Any breach or unavailability of the constituents of the CI and CII may have a nation-wide ripple effect, which could be economically or politically destabilizing. The security of both private and public entities is a shared responsibility and is pertinent to the overall security of the CII. The whole idea of PPP is to set appropriate requirements and expectations, and then align them to get the desired results.

PPP in Cybersecurity: The Existing Landscape

Cybersecurity requires joint efforts and collaboration at both the strategic and operational levels. At the strategic level, intense and close discussions are desirable to set the agenda, and outline the objectives and deliverables. Operational coordination helps in executing the objectives set for the specific segments, solutions, products or prototypes, in a time bound manner. Moreover, no single entity—whether an organisation or a nation state—can muster the requisite investment, capability, or technology which are required to strengthen cybersecurity. It is broadly understood that coherent and concerted efforts need to be undertaken by both governments and industry to evolve an ecosystem where all the stakeholders can make contributions towards the common goal of achieving the security of the cyber and information systems or infrastructure.[3]

Following extensive discussions with private sector representatives, a Joint Working Group (JWG) was established under the chairpersonship of the Deputy National Security Advisor in 2012. With representation from the government and the private sector, the JWG had constituted five Sub-Groups: for setting up of Information Sharing and Analysis Centres (ISACs) in critical sectors like Banking, Telecommunications and Power; establishment of Centres of Excellence (CoEs) on technology and policy research, Standards, Audit; Capacity building for law enforcement agencies and cyber forensics; and establishment of testing laboratories for telecom and IT equipment.[4] The JWG was established with the guiding principles and objectives of promoting the convergence of efforts of the public and private domains, leveraging existing institutions and creating new ones, implementing PPP, building policy and legal frameworks to ensure compliance, and to establish India as a global hub for the development of cybersecurity products, services, and human resources.

The JWG had also charted out a four Point Roadmap for PPP on cybersecurity issues, namely: institutional framework, capacity building, security standards and audit, testing and certification. The roadmap envisioned PPP as the medium to bridge the capacity gap through education and training, build a competency framework for skills assessment and certification, generate awareness, and fund research and development for indigenous cybersecurity products.[5] As per the recommendations of the JWG, the Joint Committee on International Cooperation and Advocacy (JCICA), as a permanent advisory committee of the JWG, was established to promote India's national interests at various international platforms on cybersecurity issues. To enhance the level of preparedness and assurance in cybersecurity, special focus was given to security standards, audit, and guidelines for the acquisition of IT products and services. The private sector was envisaged to be an active partner in defining baseline and enhanced security standards for critical sector organization and acquisition of IT products. Also, in order to address the growing concerns related to supply-chain vulnerability, the JWG solicited private sector partnership to establish national testing and certification schemes, build competence for the manpower thereof, and to set up private owned accredited testing labs.[6] Over the last five years, PPP has been a cornerstone of India's efforts, as gradual progress is being made to enhance capabilities and capacities.

The JWG is now chaired by the National Cyber Security Coordinator. The roadmap laid out by the JWG had also called for the establishment of Information Sharing and Analysis Centres (ISACs) in various sectors. ISAC for the power sector is already operational under the Central Electricity Authority.[7] Similarly, the banking sector established Indian Banks: Centre for Analysis of Risks and Threats (IB-CART) in March 2014,[8] under the Institute for Development and Research in Banking Technology. India was accepted as Common Criteria (Common Criteria for Information Technology Security Evaluation) Certificate Authorizing Nation in 2013,[9] functioning under the Standardisation Testing and Quality Certification (STQC) Directorate of the Ministry of Electronics and Information Technology. The Joint Working Group had also called for multi-disciplinary Centres of Excellence in Cybersecurity areas, in best practices, forensics, cyber crime investigation, studies, and technical as well as policy research. Efforts are already underway to meet these objectives.

However, it is increasingly challenging to identify synergies and build partnerships accordingly so that the gains are mutual for the government and the private sector. The Indian case, its requirements and experiences, are different from any other country; so comparison with the prevailing models—for example in the USA, the UK or Europe—may not be relevant. The broader question, nevertheless, is to strike the right balance between objectives and deliverables to make PPP operational in the Indian scenario.

Making PPP Operational: Finding Synergies

Every PPP model is sector and country specific. Due to varying technology requirements and the maturity of the private sector in a country, PPP models are precisely designed to suit the needs of the prevailing time. In the case of India, the PPP model for cybersecurity has been envisaged in the broader context of the growing technological prowess of the private sector in building products and delivering services related to cybersecurity, considering the influx of both domestic and foreign multinationals. Recognising the necessity of the private sector in this endeavour, the government is shaping policy frameworks in the corresponding areas where the existing synergies could be leveraged for optimal benefits. The further effort is to inculcate these synergies in the areas of common interests, both for the private and public sectors. From the point of view of cybersecurity, these areas range from capacity building to technology development. In line with the objectives of the National Cyber Security Policy and JWG recommendations, the following could be the key areas where PPP based models may find relevance as well as the complementarities of requirements and resources.

Research and Development

Cybersecurity is a research and development intensive discipline. As technology evolves, it would require cross-domain expertise to develop futuristic security solutions. Niche areas of technology, like Big Data and Artificial Intelligence, are already having an impact on cybersecurity. Even breakthroughs in the disciplines of quantum computing and cryptography have direct implications for information security. However, enabling R&D requires the intensive participation of private industry to either support such projects or ideas in the academic institutions or to transform the prototypes from laboratories into products and solutions of the highest standards, which find traction in the global markets. The role of governments is quite pertinent, particularly in empowering academic institutions, funding their research projects, and in facilitating the private sector through incentives for investments in R&D. PPP is the established means to nurture an innovation ecosystem.

To do so in India, the private sector has to step-up in not just by funding research projects or laboratories in universities, but to integrate the research throughput with their own product lines and services supply chain. As India attracts international technology players to establish their R&D centres in India,[10] cybersecurity could be one of the prime areas of focus.

Best Practices, Standardisation and Testing

The very process of developing Best Practices or Standards is collaborative, where the private and the public sector share equal responsibilities, given their stakes and investments in information systems. Devising national level practices for both, to protect cyber assets and the Critical Information Infrastructure, warrants representation and participation from both the public and private sectors on an equal footing. In general, cybersecurity encapsulates diverse entities: from service providers for Internet and telecommunications to technology integrators, or from civil society to government departments and law enforcement agencies.

Rather than being enforced, best practices work better if the whole process is consultative since the beginning, and encompasses the interests and experiences of all the stakeholders. Private enterprises can also play a pivotal role in enhancing the testing and certification facilities for IT and telecom products as part of the Common Criteria Certification Scheme. Duly accredited laboratories under the private sector, adhering to the highest quality standards, would be an important step towards reducing supply-chain vulnerabilities. For this arrangement to be productive, the model has to be economically viable for the private entities owning the laboratories. The government can at best facilitate the installation of these laboratories, and ensure quality control.

Technology Development

Technology remains the cornerstone of cybersecurity, despite the unprecedented attention attracted by the policy initiatives and strategic thinking in the discipline. The very process of technology development, from conceptualization to product deployment or service delivery, is fast-paced, transnational, and a majority of it happens outside the governmental system. No single entity, whether it is a government department or an industrial house, can afford to develop the desired technology on its own. The resources, skill-sets, investments, knowledge, and wherewithal are wide-spread. The process, therefore, involves a wide cross-section of actors, human resources, investors, partners, suppliers, and vendors, working across different time-zones in a highly competitive environment.

In such a dynamic innovation ecosystem, private players are definitely at an advantage, with an edge over technology development. Governments must leverage this expertise, both as a consumer of superior technology and security products and as a net security provider to the state. On an equal footing, the government and the private sector can muster resources and invest in technology development to tackle emerging challenges. This allows private entities to develop their products and service offerings in a competitive international market, while the government can facilitate the industry either with establishing laboratories and start-ups, or building a conducive market place which is lucrative for private players.

Regulatory Role

Governments also have to be the regulator of the markets and business sectors to ensure policy implementation and compliance. Given the prominence of the private sector in the domestic technology market, regulatory roles could be shared among the public and private sectors. This could possibly be in the shape of self-regulating or self-coordinating bodies, akin to the Sector Coordinating Councils in the USA, as discussed in Chapter 5. On similar lines, India's IT and ITES sectors have a well-functioning, self-regulation mechanism to manage cybersecurity risks. Owing to the rising importance of data security and privacy as a global phenomenon, the National Association of Software and Services Companies (NASSOM) established a self-regulatory organization, the Data Security Council of India (DSCI), in 2008. The DSCI was established to focus exclusively on policy and management issues related to data protection, and to ensure that the data security practices of the IT and ITES industry in India are synchronized with international standards. Spearheaded by NASSCOM, India's software industry is a prime example of how effective self-regulatory bodies can be for the implementation of cybersecurity policies and devising legislations related to cybersecurity.[11] This sharing of responsibility in the form of self-governed, self-regulated, or self-organized bodies in an open and participative environment could be more productive than traditional stringent regulatory-led mechanisms.

Training and Capacity Building

The discipline of cybersecurity is facing an acute shortage of human resources. This pertains not just to highly-skilled technical expertise, but also in areas which have had a significant impact due to the increased use of information technology, such as crime investigation, digital forensics, prosecution as well as day-to-day network and information security management. With such a constrained supply of professionals and well-trained human resources to execute their traditional jobs under the new environment, the PPP could be of immense help to the government. Industry and academia can develop and impart focused training programs, both at the graduate or post graduate levels as well as for working professionals who need capacity building to dispense their jobs. A successful example in India is the Cyber Labs Programme of DSCI, which played a vital role in augmenting the existing infrastructure of forensics labs in Mumbai, Bengaluru, Pune, and Kolkata for the law enforcement agencies.

Training and capacity building could be the first step towards a partnership-based approach to cybersecurity in India. As a low-hanging fruit, partnering with private entities across the industry and academia, the Ministry of Human Resource Development can target to bridge the widening gap in skill-set requirements. With private education already prevalent in India, graduate and post-graduate courses in information and cybersecurity courses, such as engineering, MCA, M.Tech, Ph.D. or MBA could bridge the prospective human resources gap. Alongside, working professionals can also leverage these institutions for short courses to upgrade or enhance their skill-sets and professional knowledge.

To especially tackle the rising cases of cyber crime and develop a deterrent against them, the Ministry of Home Affairs and the Ministry of Law and Justice can utilize training infrastructure developed in partnership with the private sector to build the capacity of the law enforcement agencies in digital forensics and criminal investigation, as well as strengthen the judicial process with prosecutors and judiciary trained for adjudicating cyber crime. The continuous need for training and education of professionals and workers, whether specialised or for general awareness in cyber, information, and network security, could effectively be met with PPP based educational models, mutually beneficial for the government and the private sector.

Building Deterrence: Cyber Threat Information Sharing

As discussed in Chapter 2, building a deterrent against threat actors is a massive effort, practiced either through denial or punishment. Threat actors, whether state or non-state, have gained technical superiority and sensitive information across governmental databases—the intellectual property of private entities or key industrial functions, are all at persistent risk. Governments and private entities have multiple partnership avenues, which could explore building strong technology enabled defences to deny hostile activities and deter at least the lower spectrum of threat actors; or to share threat information for situational awareness, threat detection, attribution and, if applicable, retaliatory action against the perpetrators. The private sector has played a constructive role, whether it was the Sony Pictures hack in 2014, the 2016 Democratic National Committee email leak investigations, or the Wannacry ransomware attack in 2017. The technical inputs and analysis lent by private entities have been critical to the respective governments in their investigations and for garnering support for the public attribution of cyber attacks; eventually strengthening their cyber deterrent practices. The avenues for the public and private sector partnerships are abound, whether in strategic thinking, norms development, internet governance, or standards development.

The key areas where PPP could flourish and overcome the impediments for cybersecurity are probably common and apparent for most nation states; but the operational challenges have restricted their success to a larger extent. As the first step, public and private sectors have to break their silos, and move forward from traditional approaches to problem solving in the case of cybersecurity. To think of the private sector merely as a service provider, a system integrator, or a product supplier, and the government as just a regulator is a serious impediment in harnessing the true benefits of PPP. Governments also struggle with nascent areas like cybersecurity where regulatory structures are absent, and there is no widely applicable template to develop policy frameworks, practices, and regulatory affairs. For governments, this is a disruptive change as they have to make and implement rules of the road while the requisite capacities to do so are intensely sparse.[12] Such constraints are further amplified in developing countries where public administration is relatively weak and the technical competence of government institutions is inadequate in the face of the rising sophistication of cyber based attacks, crimes, and hostilities. The basic idea of PPP is to leverage the different strengths, expertise, and experiences spread across the public and private sectors for the complementary roles of both in meeting developmental and social needs.

Executing PPP in Cybersecurity: Challenges and Opportunities for India

PPP as a way of collective problem solving, especially in critical infrastructure protection and cybersecurity, are well recognized and integrated with respective strategies in the USA and Europe. For the USA, PPP has been part of its critical infrastructure protection endeavours over the last two decades. The European Union has also made serious efforts to promote the PPP approach through dialogue, building understanding, and motivation. However, an exact transposition of any of the models or a direct comparison may be inappropriate. The same models in their respective contexts, political and economic systems may be a misfit for the Indian requirement.

Building trust and confidence among the stakeholders, whether they are public-private, private-private, or public-public, has been considered to be the foremost challenge,[13] and the same is applicable to India. Information sharing and knowledge exchange matures over trusted platforms, and it requires efforts from both the public and private sectors to inculcate such platforms. In order to make investments in cybersecurity sustainable for the private sector—which could be commercially uneconomic—the government can facilitate the industry with incentive schemes akin to those in the sectors with social impact,[14] such as transportation, infrastructure, energy, and electricity distribution. Relying completely on market forces to generate revenue may not be economically viable, as the private sector needs a value proposition for its investments. Moreover, fundamental differences in the outlook of the public and private sectors towards cybersecurity are a major impediment to a PPP based approach. The state, essentially the government, is the custodian of security affairs. The state apparatus tends to put cybersecurity under the auspices of national security, which basically has a direct bearing on economic and social well-being; for the private sector it is a financial and reputational risk, weighed in the cost-benefit calculus.[15] The private sector is averse to responsibilities of national security matters, and it has to prioritise the interests of the shareholders. The government, at the other end, has a responsibility towards its citizens. Aligning the interests and perspectives of both these sectors is essential to chart out common objectives.

Cybersecurity is one of the areas where governments can achieve much more if PPP based models are implemented in the right spirit. One such initiative in this regard has been the India Smart Grid Forum (ISGF), a PPP initiative of the Ministry of Power for the accelerated development of smart grid technologies in the Indian power sector. The forum provides a platform for public and private stakeholders, research, and power utilities to exchange ideas and information on smart grids.[16] The 10th Working Group of the ISGF is aligned towards the objective of enhancing cybersecurity in power systems.

As a paradigm shift, the PPP is more about partnership which matures over time, and goes beyond a buyer-supplier relationship. One interviewee from the power sector explicitly underscored the need for a mechanism where the executives in the government may directly liaison with private entities to find technical solutions for the problems they face on a day-to-day basis. Such loose arrangements, without the tedious process of tenders and bidding, may be much more effective, fast and productive for executives engrossed in their daily activities. India is also slated to play an important role in international norms development for cyberspace and Internet governance. One of the speakers from the private sector, at the workshop on "Geopolitics of Cyberspace: Creating Space for India" held at the IDSA, drew attention to the pressing need for private industry in India to come together with government and civil society to work out the modalities and frameworks within which norms could be established to ensure a healthy cyber environment. To enable such an environment for the candid exchange of ideas, the private and public sectors have to set their expectations right.

At the workshop on "Critical Information Infrastructure: Securing the Power Sector" held at the IDSA, experts responding to the survey question posed regarding expectations from the government, unanimously stated that the government should perform the dual role of supporting as well as regulating cybersecurity activities. On the question pertaining to Public-Private Partnership initiatives, the experts stated that the partnerships should be built around auditing, training, the establishment of test beds and laboratories, and research and development to aid in the indigenization of cybersecurity products.

In a broader perspective, for PPP to flourish in India, efforts should be concentrated in three areas.

- *Operations*: Private sector engagement in day-to-day business operations related to cybersecurity, with readily available or custom-made technology solutions. Being different from the traditional approach, the private sector should be made a partner in the first step of requirement analysis and solution development. The technology provider can better understand the context in which the technology solution has to be implemented, evaluated, and accordingly modified. This can also ensure more responsibility and accountability to private sector entities as technology integrators. Testing laboratories, certification facilities, test beds, and self-regulation, etc. are key operational areas for building long-term voluntary industry-government partnerships.
- *Technology Education and Research*: This domain certainly needs focus and some novel approaches to inculcate technology research in cybersecurity specific to the needs of the sector, be it energy, communications, banking or transportation. Institutions for technology education can build their respective competencies in either of the verticals of cybersecurity or distinct to the needs of the industrial sectors in partnership with the public and private sectors. Educational institutions can, therefore, supplement the training and capacity building needs of these sectors. Technology research should encompass industry leaders as well as start-ups. Deeper engagement amongst the private players can offer the right kind of expertise for a specific problem, which may also lead to out of the box thinking. This particular approach would also need changes in the existing organisational processes, with collaboration being driven by the requirements and research interests rather than pricing and contracting. For the cross-fertilisation of ideas, the employees and executives from the public sector should also be provisioned to engage in research activities, across both academia and industry. The private sector has to take the lead in this segment, and make investments in research activities a worthy business case.
- *Policy Research*: At the apex of policy making, high-quality multi-disciplinary research can help policy makers devise optimum solutions and policy frameworks. Therefore, research organisations with cross-sector expertise, drawing in resources and expertise both from the

government and the private sector, are necessary to feed into the policy making apparatus. For such bodies to be constructive, the government has to facilitate them with information, inputs and interactions.

The private sector must not just lend its expertise in technology solutions, but rather forge a long term partnership to harness real benefits. In this context, PPP can truly help public sector entities in building their own capacity in cybersecurity practices, deploy the best available solutions and, at the same time, engage private players in finding solutions to the problems which may not necessarily have a readily available solution. Technology advancements bring disruptive changes in existing organisational practices and cultures, and cybersecurity is apparently quite close to it. Traditional approaches are not suitable for non-traditional security threats to the society, economy, and intellectual property, or even to the critical infrastructure. The stakes, whether financial or reputational for the private sector or social or economic for the public sector, are quite high. Going forward, PPP as a practice would be a departure from the existing model of engagement, which is primarily based on contracting and outsourcing. India has to find the right balance of regulatory and voluntary initiatives to align the interests of the government, market, and the society. For PPP to flourish in a real sense, public and private sector entities should concentrate their energies and engagements across operations, technology research, and policy research.

NOTES

1. "Power in India Executive Seminar Series", Ryerson University and Tata Power DDL, 18–20 September 2017, at https://www.ryerson.ca/content/dam/cue/education/TataPower ExecutiveSeriesOverviewFinal.pdf, p. 2.
2. "Digital India", at http://www.digitalindia.gov.in/, accessed on 08 July 2018.
3. "Recommendations of Joint Working Group on Cyber Security", Ministry of Communications: Government of India, 16 October, 2012, at http://pib.nic.in/newsite/printrelease.aspx?relid=88442, accessed on 05 April 2018.
4. "Cyber Security", Press Information Bureau: Government of India, 20 March 2018, at http://pib.nic.in/newsite/PrintRelease.aspx?relid=177722, accessed on 08 July 2018.
5. Recommendations of Joint Working Group on Engagement With Private Sector on Cyber Security, National Security Council Secretariat, National Security Council Secretariat, p. 2.
6. Ibid, p. 3.
7. Central Electricity Authority, "Power Sector: Information Sharing and Analysis Centre", at http://www.cea.nic.in/isacpower.html, accessed on 08 July 2018.
8. "Indian Banks: Centre for Analysis of Risks and Threats (IB-CART)", Institute for Development & Research in Banking Technology, at http://www.idrbt.ac.in/ib-cart.html, accessed on 08 July 2018.

9. "Indian Common Criteria Certification Scheme (IC3S)", at http://www.commoncriteria-india.gov.in/, accessed on 08 July 2018.
10. Raja M. Mitra, "India's Emergence as a Global R&D Centre", *Swedish Institute for Growth Policy Studies*, Working paper R2007:012, pp. 7–8.
11. Nir Kshetri, "India's Cybersecurity Landscape: The Roles of Private Sector and Public-Private Partnership", *IEEE Security & Privacy*, Vol. 13, no. 3, pp. 16–23,
12. Myriam Dunn Cavelty and Manuel Suter, "Public-Private Partnerships are no silver bullet: An expanded governance model for Critical Infrastructure Protection", *International Journal of Critical Infrastructure Protection*, Vol. 4, No. 2, 2009, pp. 179–187.
13. "Public Private Partnerships: Cooperative models", European Union Agency for Network and Information Security, November 2017, p. 5.
14. "Public-Private Partnership", Internet Security Alliance, at https://isalliance.org/policy-advocacy/public-private-partnership/, accessed on 05 July 2018.
15. Madeline Carr, "Public-private partnerships in national cyber-security strategies", *International Affairs*, Vol. 92, No. 1, 2016, pp. 4362, at https://www.chathamhouse.org/sites/default/files/publications/ia/INTA92_1_03_Carr.pdf, p. 55.
16. "About us", India Smart Grid Knowledge Portal, India Smart Grid Forum, at http://www.indiasmartgrid.org/about-us.php, accessed on 08 April 2018.

CHAPTER 8

India's Strategic Options in a Changing Cyberspace

A constantly mutating and changing cyberspace in terms of opportunities, threats, technologies, and utility, calls for nimble responses on the part of policy makers around the world. By way of anecdotal testimony, when this research project began, the focus of the global community was on the norm building process, with the end goal being of an open, secure, and global cyberspace. In the space of the succeeding two years, the global consensus has more or less collapsed, the battle lines based on alternate ideologies and perspectives have become more deeply entrenched, and hitherto disruptive attacks have transitioned into more destructive attacks, crossing many redlines in the process. Cyber attacks and exploitation methods have become more sophisticated and even more difficult to defend against, with old forms of attacks (such as ransomware) mutating from being merely acts of cyber crime to state-sponsored attacks against which effective remedies are yet to be found. In cyberspace, the perception that it is now essentially "each man for himself and the devil take the hindmost" has taken root and led to a change in priorities; also reflected in this book with norm creation being considered as a subset of cyber deterrence.

The idea of cyberspace as a battlefield for strategic dominance is yet to be internalized by most policymakers and even information security experts, both of whom are still most concerned with the mitigation of risks from cyber attacks and the remediation thereof, rather than deterring such attacks in the first place. While the former is important, equal attention needs to be paid to the latter. Cyber defences, while thwarting a wide variety and majority of the attacks, fall short of deterring adversarial States and hostile non-state actors from waging malicious activities, in and through cyberspace. Cyber deterrence, both through the means of denial and punishment, provides a strategic framework within which a comprehensive proactive response can be charted out to dissuade cyber attacks at the first place. Cyber deterrence is the cornerstone of an effective cybersecurity strategy, and it must be practiced in full spirit, scope, and capacity. A synchronised and concerted national effort, and strong political will, is necessary for effective cyber deterrence. In essence, the practice of cyber deterrence has to interlace the roles, responsibilities, capacities, capabilities, infrastructures, and skill-set spread across the government, the armed forces, and the private sector.

India has many of the building blocks for effective cyber deterrence in place; but it is somewhat underprepared in many other aspects. If any lesson has been learnt about cybersecurity, it is that it is only as strong as the weakest link. The considered view of experts is that, from a deterrence perspective, there is considerable emphasis on deterrence by denial but not so much on deterrence by punishment. In the first instance, this would definitely entail some reflection on the role of the armed forces since, under the laws of armed conflict, they bear the primary responsibility of guarding the nation state across the various domains.

Developing the ability of deterrence by threat of punishment requires the building up and the demonstration of offensive cyber capabilities. Some countries, such as the USA and the UK have set up cyber commands, and have been quite forthright about issuing declaratory doctrines on offensive cyber operations; others have preferred a more recessed deterrence, neither confirming nor denying capabilities.

For the offensive aspects of deterrence to be effective, attribution capabilities also have to be developed substantially, using both technical means and analytical models. Higher confidence in attribution can justify punishment, and strengthen deterrent capability by setting a precedence that threat actors, including nation states, have to pay the price for any act of hostility. This also breaks the vicious circle of eroding deterrence due to uncertain and inaccurate attribution. Going forward, the ability to successfully attribute a cyber attack with higher probability and conviction will justify offensive actions and underpin cyber deterrence capability. India should focus on investing in the technology, intelligence gathering infrastructure, intelligence sharing platforms, and bilateral relations to augment the credibility of its attribution.

Cyber deterrence as a framework has to be pursued with care since the dangers of misperception and over-reaction resulting in cascading effects like a cyber arms-race and instability are quite high. Therefore, international cooperation on norms building provides both an opportunity for confidence building as well as signalling, as it has for other strategic domains like nuclear and space. Regional organizations and economic groupings, such as ASEAN and OECD, have continued to carry out the work of fleshing out and implementing many of the norms contained in the UNGGE reports. However, unless the UNGGE or a similar apex process is reinstated in the near future, these efforts would falter over a period of time. It is also imperative to have a cybersecurity forum operating in the South Asian region not just because the importance of regional fora are likely to increase over time, but also because the experience of other regional fora show that they play an important role in regional cybersecurity.

An analysis of the various forums shows that fragmentation is increasingly a liability, with some forums subject to capturing by the vested interests of certain States, while others have been initiated by parties with the express intent of propagating particular approaches and perspectives. Even if many of these forums have been around for many years, their largely *ad-hoc* nature and lack of sustained funding reduces their effectiveness, and makes them susceptible to State capture while other states go forum shopping to find fora that best represents their interests. This is even truer for multi-stakeholder forums which are consequently losing their relevance.

It goes without saying that it is in the interest of all countries to have a multi-pronged approach to cybersecurity. Whilst they are well within their sovereign rights to undertake all measures possible to deter attacks, it is also prudent on their part to cooperate and engage constructively to ensure that the global information infrastructure remains stable. As seen in Chapter 3, India's participation in the global conversations on cybersecurity has encompassed all mechanisms, from multi-stakeholder to multilateral, and from regional to bilateral. However, it has defaulted to playing the role of a bridging power or honest broker and, in the process, refrained from evolving definitive positions on many of the issues. With the erosion of global consensus on the future of cyberspace, India should partner with like-minded countries such as

Germany, the Netherlands, South Korea, and Japan to push the norm making process forward.

At present, India has bilateral strategic cyber dialogues with over 10 countries, and engagements with multilateral organisations such as the EU and ASEAN; it has also signed myriad cyber MoUs, agreements and joint statements with a further 40-odd countries. Even if the intent is to further cooperation, much of the content of these MoUs are boilerplate resolutions to share best practices, exchange information on cyber threats, strengthen IT infrastructure, intelligence sharing, etc. Going forward, there is a need to have more cooperation with other developing countries if India is to build up a constituency of like-minded countries. This cooperation should leverage India's existing strengths in the field of IT and cybersecurity. Existing MoUs with developing countries like Mongolia, Bangladesh, Bhutan, and Tunisia focus on areas like strengthening cooperation in e-Governance, m-Governance, and e-Public Services Delivery, all of which also have a cybersecurity angle to them. The immediate neighbourhood could also be a good starting point for extending India's expertise in executing large e-Governance and IT infrastructure development projects, to lay the foundation of long term cooperation over cybersecurity. Deepening cooperation on cybersecurity for member countries from the Bay of Bengal Initiative for Multi-Sectoral Technical and Economic Cooperation (BIMSTEC), as proposed at the meeting of the BIMSTEC national security chiefs, is a welcome step in this direction. India should take every necessary step to sustain this momentum, and institutionalise such workshops and experience sharing platforms in the long term.

India has a lot to learn from the global experiences in Critical Information Infrastructure Protection, which have matured over the years with close support from the private sector. Self-organized, self-run, and self-governed private sector initiatives can play an important role in formulating a better working and mutually beneficial relationship among the private and public sector entities. They can facilitate discussions and also ensure the representation of operators of critical infrastructure entities in the policy making process. Cross-sector coordination is vital for India to resolve the issues pertaining to interdependencies which cut across different sectors, but are not being discussed in sufficient detail. As recourse to the general concern of enterprises from the critical infrastructure sectors, the nodal agency NCIIPC may consider periodic interaction with the board of directors of these enterprises to underscore the importance of CII protection and cybersecurity in general, against the backdrop of national security. The government, particularly in this case, has to support or facilitate the private entities through incentives as well as ensure compliance through regulation. However, government alone cannot achieve these disparate objectives, whether it is in CIIP or in the practices of cyber deterrence, or Active Cyber Defence, going forward.

An essential element in collaboration is working out a mechanism to bring the private sector into the process, not just in execution but as a strategic partner. Active Cyber Defence, as it pertains to the private sector represents one extreme, and could be seen as an inevitable outcome if States do not adequately shoulder the responsibility of providing security. However, Active Cyber Defence as a concept could be problematic since the laws of most countries do not allow private entities to undertake even the most basic actions to respond to cyber attacks.

The analysis of Active Cyber Defence serves also to highlight the role the private sector can play, to a greater or lesser extent, in partnering with the State and lowering its burden in a domain which is unique in having such a wide attack surface. Whilst Public-Private Partnerships are being implemented at the country and the State level, they are yet to take off at a supranational or regional level. With the coming onslaught of the Internet of Things, where as many as 31 billion devices are expected to come online by 2020, it will require active partnerships between the government and the private sector to keep this burgeoning domain secure. Private sector involvement is a prime requisite to achieve the objectives laid out in the National Cyber Security Policy, pertaining to R&D, capacity building, standardisation and testing, or even to building deterrence against threats in cyberspace.

The private sector, by virtue of its placement within the economy and the ability to muster resources, has to be an integral part of India's quest in securing cyberspace. The cybersecurity ecosystem in India, as seen in figure 8.1, is still at a nascent stage and as a consequence, too siloed and government centric. Public-Private Partnerships and breaking the silos can be a true value proposition if the private sector can provide technology and expertise for building defences, desirable offensive capabilities; aid in crime investigation and law enforcement, bridge the human resources deficit; build innovative products and solutions; fund university research programmes; and establish laboratories for product testing and certification, just to name a few. India is

Figure 8.1: India's Cybersecurity Ecosystem – Visual Representation#

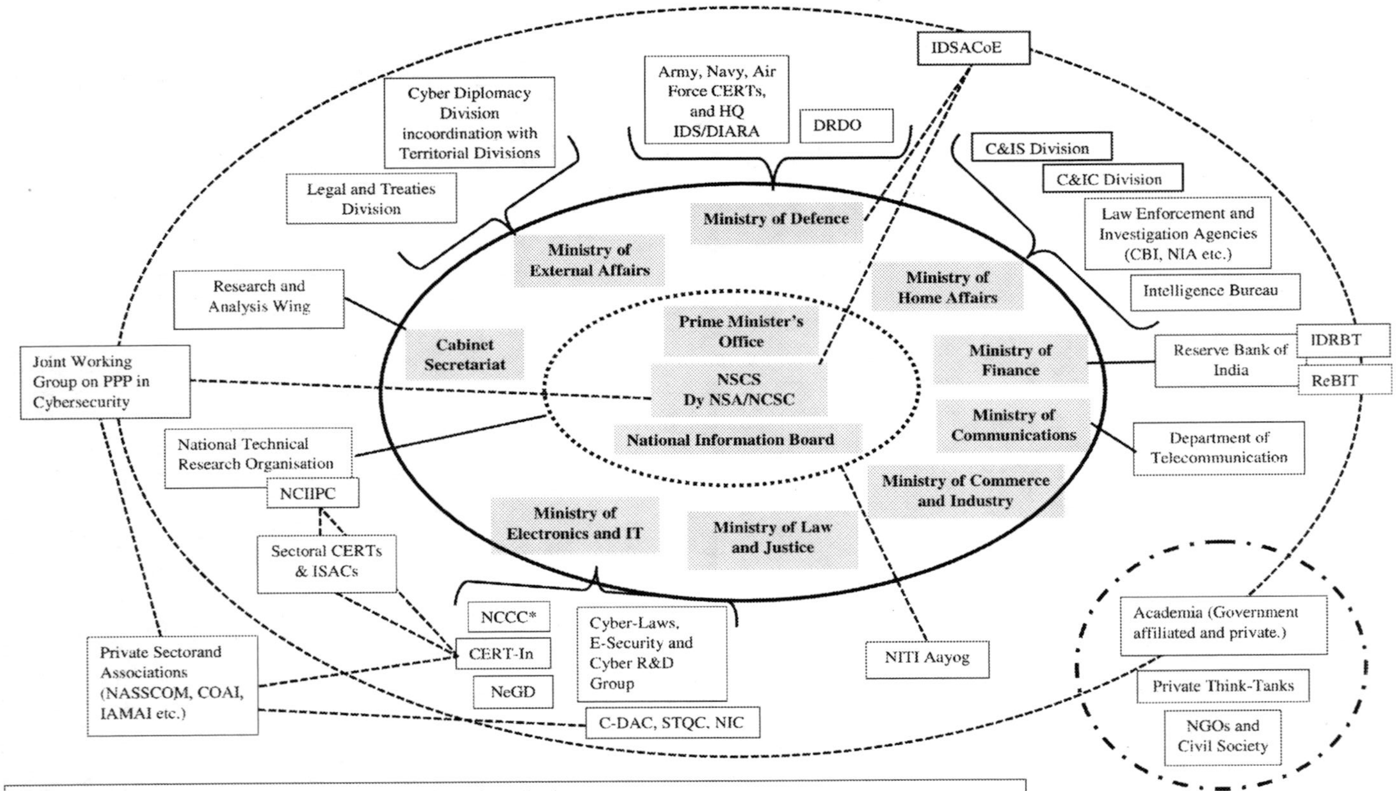

Inner circle: Apex bodies for cybersecurity policy formulation and coordination.
Middle circle: Nodal Ministries, responsible for overall supervision and coordination of cybersecurity policy making.
Outer Circle: Departments and entities mandated for cybersecurity policy formulation, policy inputs or implementation, and their interactions.

#Based on authors' interpretation of information and documents available in the public domain.

in dire need of indigenous production and manufacturing. This aspect has been flagged and underscored at various forums, also finding a mention in the National Cyber Security Policy as well as in the recommendations of the Joint Working Group on cybersecurity. Whilst 100 percent indigenisation would be unrealistic, these efforts have to go hand-in-hand with stringent auditing and regulatory mechanisms to ensure the continued integrity of systems, processes, and the supply chains.

As India is poised to make strides as a digital and knowledge based economy, the technology and political issues surrounding encryption are also quite important for it to have a practical and viable policy. Encryption ensures the security of sensitive data and information; it is also the bedrock of a vibrant digital economy. Weak policy or regulatory measures or even inferior technology can seriously undermine growth prospects. India has to be an innovator and an early technology adaptor to sustain the growth momentum in the services and high-technology segment. Trust and confidence of enterprises as well as consumers in India's promising technology development sector are pertinent, and must be strengthened further. With intensifying global competition for technology, the disciplines of quantum computing, quantum cryptography, and artificial intelligence are poised to throw open immense business and economic opportunities, with clear implications for information security and cybersecurity.

As noted in Chapter 6, these technologies hold the potential of disrupting existing practices and approaches to cybersecurity. They can even radically change the way information is protected and secured communications carried out. However, India fares average in the surging technological competition for encryption, block chain, AI, and quantum information sciences. India has rolled out a discussion paper and established a task force to work out the modalities and adapt it for the social good, with priority given to healthcare, agriculture, and transportation sectors. The discussion paper makes a case to maximise the late-movers' advantage. This notwithstanding, India has to graduate from 'late-majority' to an 'early adaptor' or an 'innovator' on the technology adoption lifecycle curve. Possibly, more investments on the part of both the public and private sectors, in experimental facilities can help the theoretical research community to move their ideas from the laboratory to prototypes or transform them as finished products. As multi-disciplinary research areas have strategic importance for economic and social development, integrating expertise from the diverse domains of computer science, physics, electronics, materials, data science, information science, and mathematics is vital to succeed in these areas where other countries are also striving for technology dominance. India has to contemplate and respond to global advancements in order to develop a research oriented innovative ecosystem, harnessing technologies for cybersecurity, and it should not end up as a mere consumer in this emerging and potentially expansive market.

In the final analysis, cybersecurity has entered a new era of uncertainty. The old cyber order, which was predicated on the principles of a global, open, and secure cyberspace, has failed on many fronts since major countries are unwilling to practice the strategic restraint necessary to continue the trust based architecture on which cyberspace has been built. Thus, of the various themes and issues covered in this book, it may be safely expected that issues such as deterrence and Active Cyber Defence will prevail over attempts to create a norm-based cyber order. The failure of the latter was only inevitable, as major countries have come to see such norms as obstacles in the race to dominate cyberspace, be it through technology dominance or developing espionage, cyber warfare and other capabilities, which are as yet unknown.

It is imperative to take cognizance of the new directions in cyberspace policy making around the world, and the impact it is having on cybersecurity. This sea change in thinking must be addressed through appropriate national policies. There must be a realistic assessment of successes and shortcomings in crucial areas such as critical information infrastructure protection and the role India has played in the norms building process. Public-Private Partnership is another critical area which is in dire need of evaluation, the many task forces that have gone into it, notwithstanding. Whilst many of the issues have been taken up and policies and decisions announced, and this holds true not just for the Public-Private Partnership, but most of the other issue areas concerned, insufficient energy has gone into implementing these decisions. The success or failure of India as a major power in cyberspace will largely be shaped by its strategic thinking, ability to provide thought leadership, interactions among multiple stakeholders and certainly, the efficacy of mechanisms devised for cyber policy and its implementation.

Recommendations

During the course of research and writing of the book, the authors consulted and held discussions with experts, attended conferences, roundtables and seminars, and carried out a comprehensive survey of literature. The following recommendations have been culled out of these interactions.

(1) India should not be oblivious to the emerging landscape of deterrence in cyberspace. India should not refrain from developing and being overt about the use of offensive cyber operations to deter threats in cyber as well as in other domains where India maintains superiority.

(2) For deterrence to work effectively, India has to establish the credibility of its threat by communicating clearly the capabilities India possesses and the intent to use them in response to hostile acts.

(3) A Tri-services Cyber Command is desirable, but a new or existing agency mandated to execute offensive operations in cyberspace should be established at the earliest. The agency should function within the purview of the armed forces, staffed with human resources from the government and private sector. For strategic decision making, it may replicate or draw inferences from the Nuclear Command Authority. The agency should have provisions for lateral movement of expertise across the government, armed forces and private sector. The territorial-army model may be considered to bring in expertise from the private sector.

(4) India should start contemplating on "Cross-domain" deterrence. A task force on the subject can fuse the aspects of academic and operational thinking to weigh the benefits and risks of such an approach to deterrence, specifically from an Indian perspective.

(5) Active Cyber Defence, despite the controversial elements such as "hacking back", should not be discounted entirely. Both the

governments and the private sector, as responders have to elevate their level of preparedness.

(6) Remote intelligence gathering, an integral part of ACD is tantamount to hacking and therefore illegal under prevailing National laws. To strengthen ACD measures, the rule of engagement must be laid down in accordance with the IT Act, 2000 for a government agency or a private entity in India.

(7) Private entities and law enforcement agencies will necessarily have to work in tandem to develop the necessary skills such as threat intelligence, domain awareness and technical expertise, and it is possible if the government facilitates the process.

(8) Cybersecurity policies of critical infrastructure sectors (notified as protected entity under the IT Act), should be top driven, preferably as a board agenda, so that policy implementation becomes administratively easier at the CISO level. An annual plan for cybersecurity, with the involvement of all key departments (OT-HR-Finance-Admin-Safety-Legal etc.) and its regular review and audit would be effective in managing cybersecurity policies and practices at the organisational level. Cybersecurity should not be perceived as purely an IT activity, and the core team for execution of cybersecurity policies should be composed of executives from all the key departments.

(9) Formal and informal dialogue platforms should be established for regular interaction of CISOs from organisations part of the Critical Infrastructure and other key players outside it. The nodal agencies, NCIIPC, CERT-In and sectoral CERTs may facilitate these platforms for candid exchange of ideas, processes and practices related to technology and management aspects of cybersecurity. These platforms should be made sustainable and interest driven, rather than as a one-off activity.

(10) Sectoral coordination should be supplemented with cross-sector coordination, as interdependencies across sectors have not received due attention in the Indian context. Academic exercises and simulations could be leveraged to understand the complexity of information and commodity exchange across sectors and the impact of disruptions thereof. The subject of interdependencies has received

substantial attention internationally, both for research and policy making.

(11) Incident, breach and vulnerability disclosures should be made mandatory, at least for the critical sectors. This may be backed with legal action and penalties for non-compliance.

(12) Government may mandate testing of the imported products pertaining to information systems, networking and communications, Industrial Control Systems or SCADA systems sourced to be deployed in the critical infrastructure sectors. Given the vast demand for these products in the Indian market, government may also consider mandating key foreign players in the market to establish testing and certification laboratories in India.

(13) Private sector should assist government endeavours, and the likely areas could be audits, training programs, test beds/laboratories for quality certification, and indigenization of cybersecurity products.

(14) From the view point of technology for information and cybersecurity, India needs to pay immediate attention to basic and applied research in cryptography, AI, big data analytics and quantum information sciences. These all warrant investment of capital and human resources in both fundamental and applied research. Therefore, the institutes of eminence for research in basic sciences and technology should receive adequate funding. The research projects should not duplicate efforts and independently they should all converge at the national effort to bridge the research and technology gap.

(15) NITI Aayog can drive and coordinate research initiatives, some of which are being executed or conceived under the Department of Science and Technology, Defence Research and Development Organisation or the Department of Defence Production, as part of their directed research programmes. A single entity spearheading advanced research, as a national priority, would be more effective than disparate efforts in the civilian and military domains.

(16) Addressing the specific needs of India, and keeping an eye on their regional and international utility, India should aggressively cultivate innovation culture by integrating industrial and governmental projects with academic institutions and production. India should also leverage the innovation hubs being established in India by global technology players, such as Accenture and Microsoft.

(17) India should start practicing Public-Private Partnership in true sense, involving private sector entities since the first step of requirement analysis and solution development, in essence, building long-term voluntary industry-government partnerships.

(18) As a departure from the existing model of engagement with the private sector, which is primarily based on contracting and outsourcing, government has to bring private sector in long-term planning and execution; aligning the interests of the government, market and the society so that it is mutually beneficial for all. The government may contemplate on making education, trainings and laboratories for advanced research a good investment proposition for the private sector.

(19) Given the technological superiority of India in the immediate neighbourhood, India should build leadership in cybersecurity and e-governance domains at the regional organisations such as SAARC and BIMSTEC. India should take the lead in charting out effective legal measures to curb cyber crime, and to adopt cybersecurity as an agenda item for discussions. India can establish platforms and forums for information and expertise sharing on national strategies and policies, security incidents, law enforcement cooperation, and frameworks for confidence building.

Index